DEVELOPMENTS IN FOOD MICROBIOLOGY—1

THE DEVELOPMENTS SERIES

Developments in many fields of science and technology occur at such a pace that frequently there is a long delay before information about them becomes available and usually it is inconveniently scattered among several journals.

Developments Series books overcome these disadvantages by bringing together within one cover papers dealing with the latest trends and developments in a specific field of study and publishing them within *six months* of their being written.

Many subjects are covered by the series including food science and technology, polymer science, civil and public health engineering, pressure vessels, composite materials, concrete, building science, petroleum technology, geology, etc.

Information on other titles in the series will gladly be sent on application to the publisher.

DEVELOPMENTS IN FOOD MICROBIOLOGY—1

Edited by

R. DAVIES

National College of Food Technology,
The University of Reading, Weybridge, Surrey, UK

APPLIED SCIENCE PUBLISHERS
LONDON and NEW JERSEY

APPLIED SCIENCE PUBLISHERS LTD
Ripple Road, Barking, Essex, England
APPLIED SCIENCE PUBLISHERS, INC.
Englewood, New Jersey 07631, USA

British Library Cataloguing in Publication Data

Developments in food microbiology.—1.—(The
Developments series)
1. Food—Microbiology—Periodicals
2. Food contamination—Periodicals
I. Series
576′.163′05   QR115

ISBN 0-85334-999-1

WITH 15 TABLES AND 35 ILLUSTRATIONS

© APPLIED SCIENCE PUBLISHERS LTD 1982

Printed in Great Britain by Galliard (Printers) Ltd, Great Yarmouth

PREFACE

As consumers increasingly seek convenience and novelty as well as gastronomic pleasure in the food marketplace it is essential that we continue to understand, monitor and control those microbiological factors that contribute to product wholesomeness and safety. The presence and activity of microorganisms in our food materials can be considered to delight, disgust or damage us, reflecting, respectively, their beneficial role in traditional and novel fermented foods and their negative roles in food spoilage and food-borne disease. It is the aim of this and subsequent volumes, therefore, to consider topical developments in each of these aspects of food microbiology maintaining, where possible, a balance between reviews of existing knowledge and the introduction of new concepts from current research.

In this volume two chapters deal with major developments in the microbiology of the important food commodities, meat and fish, whereas another introduces a controversial approach to the control of salmonellosis. The microbiological problems encountered by extrapolation of thermal processing to high temperatures for very short times are identified in a further chapter while another author considers the impact of current developments in molecular genetics on one of our traditional fermented food processes—namely cheesemaking. Finally, new rapid and automated methods of detecting microorganisms are considered in relation to their application to foods.

R. DAVIES

CONTENTS

LIST OF CONTRIBUTORS

CELIA A. AYRES

   Campden Food Preservation Research Association, Chipping Campden, Gloucestershire GL55 6LD, UK

K. L. BROWN

   Campden Food Preservation Research Association, Chipping Campden, Gloucestershire GL55 6LD, UK

P. A. GIBBS

   Leatherhead Food Research Association, Randalls Road, Leatherhead, Surrey KT22 7RY, UK

G. HOBBS

   Ministry of Agriculture, Fisheries and Food, Torry Research Station, P.O. Box 31, 135 Abbey Road, Aberdeen AB9 8DG, UK

W. HODGKISS

   Ministry of Agriculture, Fisheries and Food, Torry Research Station, P.O. Box 31, 135 Abbey Road, Aberdeen AB9 8DG, UK

L. L. MCKAY

   Department of Food Science and Nutrition, University of Minnesota, 1334 Eckles Avenue, St. Paul, Minnesota 55108, USA

T. A. MᴄMᴇᴇᴋɪɴ

Department of Agricultural Science, University of Tasmania, G.P.O. Box 252C, Hobart, Tasmania 7001, Australia

E. Nᴜʀᴍɪ

State Veterinary Medical Institute, P.O. Box 10368, 00100 Helsinki 10, Finland

H. Pɪᴠɴɪᴄᴋ

Bureau of Microbial Hazards, Food Directorate, Health Protection Branch, Department of National Health and Welfare, Ottawa K1A OL2, Canada

J. M. Wᴏᴏᴅ

Leatherhead Food Research Association, Randalls Road, Leatherhead, Surrey KT22 7RY, UK

Chapter 1

MICROBIAL SPOILAGE OF MEATS

T. A. McMeekin

Department of Agricultural Science, University of Tasmania, Australia

SUMMARY

Many types of psychrotrophic microorganisms contaminate meat during processing but only a fraction of the initial population has the capacity to develop under a given set of storage conditions. During aerobic storage at chill temperatures Pseudomonas *species predominate and spoilage results from the utilisation of small molecular weight compounds rather than meat proteins. Within the small molecular weight group the spoilage flora also exhibit preferential utilisation of substrates. Similarly under anaerobic conditions the dominant facultative anaerobes also show a preference for certain substrates. The spoilage potential of isolates may be assessed on sterile excised muscle and studies of this type indicate that only a fraction of the flora present produce off odours associated with spoilage. Prominent among these are volatile sulphides which are detectable organoleptically at extremely low concentrations. A characteristic of most chemical, physical and bacteriological techniques for the evaluation of spoilage is that they are diagnostic rather than predictive. The concept of temperature function integration provides an alternative which can be used to continually monitor the temperature history and remaining shelf life of aerobic, chill stored meats.*

INTRODUCTION

Flesh foods stored at chill temperatures spoil as a result of the growth and metabolism of psychrotrophic microorganisms. However, Jay[91] in a review of the mechanism and detection of microbial spoilage of meat at chill temperatures indicated that after 80 years of research the precise mechanisms by which flesh foods spoil had not been elucidated nor could general agreement be reached on the validity of any method for the prediction of spoilage. The purpose of this chapter is to examine developments in the study of the microbial spoilage of meats, with particular reference to information which has become available since the reviews of Ingram and Dainty[84] and Jay.[91]

CONTAMINATION

Contamination of flesh foods with spoilage microorganisms is an inevitable consequence of modern processing methods and some consideration of contamination is therefore pertinent to any discussion of spoilage.

The organisms of significance in the spoilage of chill stored meats are the psychrotrophs which are common inhabitants of soil, water and vegetation and may enter the processing plant on the hide, fleece, feet or feathers of the animal or be added from the water used during processing. If cleaning practices are improperly carried out, large numbers of psychrotrophs may build up on plant and equipment.

Newton et al.[138] recovered psychrotrophs from hides, structural and work surfaces within an abattoir and from carcasses and meat at all stages of processing. Increased numbers were recovered on carcasses during processing through the abattoir and a seasonal variation was noted which correlated positively with rainfall and negatively with temperature. The latter observation confirmed the data of Empey and Scott.[48]

Similar increases in psychrotrophic contamination of poultry carcasses have been noted.[99,102,184] In each instance immersion washer and spin chiller water was implicated as the major source of these organisms. Barnes,[9] Clark,[32] Notermans et al.[144] and Thomas[184] reported that the skin of carcasses was free of psychrotrophic contaminants immediately after scalding. Thomas also noted that fresh water inputs at various stages of processing normally contained < 10 psychrotrophs/ml but found $770/ml$ in ice, $> 2 \times 10^3/ml$ in immersion washer water and $> 6 \times 10^3/ml$ in chiller water. These inputs resulted in carcasses with $c.\ 5 \times 10^2$ psychrotrophs/cm^2 of skin. For all samples the psychrotrophic count represented $< 10\%$ of the viable count at 20 °C confirming the estimate of Barnes[10] which indicated that psychrotrophs varied from 1% to 10% of the 20 °C viable count.

Although the literature contains a great deal of information on numbers and types of microorganisms at various stages of meat processing and much effort has been put into reducing levels of carcass contamination, fundamental information is lacking on the mechanisms of contamination of meat surfaces. Several studies have now been reported on parameters important in the contamination of various substrates: chicken skin;[17,118,142,143] hog skin;[23] pork skin and surfaces of beef and lamb carcasses;[24] chicken and beef meat[51] and cows teat skin.[141] Electron microscopic examinations have also been made of bacteria on cows teat skin[52] and chicken skin.[119,184] Although several contentious issues have been raised by these studies, this type of work offers the possibility of a better understanding of the relationship between contaminant microorganisms and food substrates. Hence less empirical means of preventing contamination or procedures for decontamination may be developed.

PSYCHROTROPHIC MICROORGANISMS

A great number of genera are represented within this physiological group, however, only a few are of importance in meat spoilage. Gill and Newton[62] listed these as follows: *Pseudomonas, Moraxella, Acinetobacter, Lactobacillus* with *Microbacterium, Alcaligenes, Vibrio, Aeromonas* and *Arthrobacter* playing a relatively minor role. In addition, psychrotrophic enteric types have been reported as components of meat spoilage associations.[15,42,56,75,152] Extensive lists of the occurrence of various microbial genera isolated from meat and meat products have been given by Jay.[93]

A great deal of confusion surrounds the use of the terms psychrophile and psychrotroph in the literature. Morita[130] pointed out that many workers had used the term psychrophile incorrectly and restricted this terminology to organisms with an optimum temperature of 15 °C or less, a maximum of 20 °C and a minimum of 0 °C. The term psychrotroph[43] refers to those organisms which can grow at refrigeration temperatures but have temperature optima greater than 20 °C. As most of the contaminant organisms on meat carcasses are likely to be exposed to elevated temperatures at some time in their life history, it follows that the psychrotrophic group will be largely responsible for the spoilage of refrigerated meats. Accordingly, the term psychrotroph is used in this paper.

Elliott and Michener[45] in their review of 'psychrophiles' in food spoilage, noted that only the lag and logarithmic phases of growth are important as the material will have spoiled by the time the phase of decline has been reached. The effect of temperature on these two phases of the growth cycle is therefore of considerable practical significance. In the simplest terms, the lag phase will be shortest at the optimum temperature for growth of the organism and lengthens as the temperature is lowered tending towards infinity as the minimum temperature for growth is reached. In the temperature range $10°$–$30°C$ where both psychrotrophs and mesophiles grow, the psychrotrophs have a shorter lag time.

Temperature also affects the rate of reproduction and generation time is particularly sensitive to slight changes in temperature in the lower ranges. The effect of temperature on growth may be measured in various ways— temperature coefficient (Q_{10}) or temperature characteristic (μ). To obtain the latter value, microbiologists have substituted growth rate in the van't Hoff–Arrhenius equation for reaction rate

$$\log_{10}\left(\frac{K_2}{K_1}\right) = \frac{\mu(T_2 - T_1)}{(2 \cdot 303 R T_1 T_2)}$$

where K_1 and K_2 are the growth rates at temperatures T_1 and $T_2(°K)$, R is the gas constant, and μ is analogous to activation energy in an enzyme catalysed chemical reaction.

Some workers have suggested that μ could be used to determine the temperature of growth characteristics of an organism, e.g. Ingraham[82] suggested that psychrotrophs should have a lower μ than mesophiles but this concept was later refuted by Haight and Morita,[69] Shaw[169] and Hanus and Morita.[71] The latter authors used several closely related vibrios including a psychrophile, a psychrotroph and a mesophile and reported μ values of 16 200, 16 400 and 14 400 cals/mol respectively. Similarly, Harder and Veldkamp[72] reported similar μ values for a psychrophile and a psychrotroph but noted a difference in the range in which the Arrhenius plots were linear (Fig. 1). Deviation from linearity occurred at $4°$–$5°C$ with the psychrotroph which indicates that a decrease of temperature below $5°C$ will result in a decrease in growth rate much greater than expected from the effect of temperature on the rate of chemical reactions in the cell alone. With the psychrophile, deviation from linearity did not occur from $+4°C$ to at least $-2°C$ and this strain would show a greater maximum growth rate at these temperatures. A small drop in temperature in the chill range will therefore have an important effect on the development of psychrotrophs which are the organisms of major concern in the spoilage of refrigerated meats.[83] Ingram and Mackey[85] also point out that the effect of temperature on microbial development cannot be considered in isolation, but will be

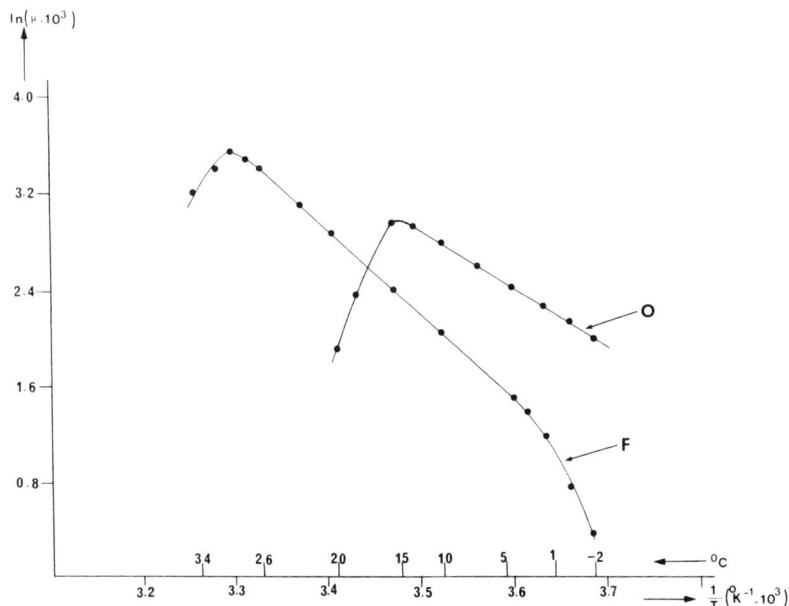

FIG. 1. Arrhenius plot of the maximum specific growth rate of a psychrophilic (O) and a psychrotrophic (F) *Pseudomonas* species at different temperatures (data of Harder and Veldkamp,[70]). μ: temperature characteristic; T: absolute temperature ($^{\circ}K$).

potentiated by other factors which, when acting together, will ultimately determine the spoilage association of a food stored under a particular set of conditions.[131] The interaction between a_w and temperature serves as a good example of this effect and is particularly noticeable when the temperature drops below $0\,^{\circ}C$ and a_w values become progressively lower.[186] Meat stored at above zero temperatures is normally spoiled by bacteria, but in the frozen state yeasts and moulds are the predominant spoilage organisms. The situation close to $0\,^{\circ}C$ is exemplified by the data of Barnes *et al.*[12] which showed an increase in the proportion of yeasts occurring on turkey carcasses stored at $-2\,^{\circ}C$ compared to carcasses stored at above $0\,^{\circ}C$. Psychrotrophic yeasts formed only a minor component of spoilage association at $0\,^{\circ}C$ and above, but were present at $> 10^7/cm^2$ of skin after 42 days storage at $-2\,^{\circ}C$. However, the a_w value of tissue ice at $-2\,^{\circ}C$ would be $0 \cdot 98$[186] and was sufficiently high to allow pseudomonads to continue growing and remain numerically dominant. Leistner and Rodel[109] quote minimum a_w values of $0 \cdot 95$–$0 \cdot 96$ for pseudomonads.

EFFECT OF STORAGE ATMOSPHERE

The spoilage association which develops on chill stored meats will be determined to a large extent by the storage atmosphere. If meat is stored aerobically then organisms of the genus *Pseudomonas* will normally predominate and this has been demonstrated for beef,[6,91] poultry[11,114,116] and porcine muscle.[56] Other psychrotrophs including members of the *Enterobacteriaceae*[15,152] and flavobacteria[116] may also contribute to spoilage if storage temperatures are allowed to rise, although some workers report pseudomonads are the major group on meat stored at all temperatures up to 15 °C.[55] However, there can be no doubt that as the storage temperature approaches 0 °C, pseudomonads represent the major group of spoilers; their competitive ability being due to faster growth rates at chill temperatures.[61]

Alteration of the gaseous environment in which meat products are stored changes the spoilage association and spoilage characteristics. This is normally achieved by enclosing the meat in gas impermeable wrapping, thereby reducing oxygen availability and increasing the amount of CO_2 in the atmosphere. The basic procedure has been used to extend the shelf life of beef,[6,70,95] lamb,[8] pork[56] and poultry.[145,176,187] The effect of the treatment is to reduce the logarithmic phase of growth and generally inhibit growth of spoilage bacteria.[93] Exclusion of oxygen and increased levels of CO_2 result in marked changes in the microflora. Strictly aerobic pseudomonads are inhibited and the new set of ecological conditions favour the development of anaerobic and facultatively anaerobic strains such as *Lactobacillus*, *Brocothrix thermosphacta* and psychrotrophic enterics. The spoilage characteristics of the product are also different from those of aerobically stored meats, which provides good circumstantial evidence for the role of pseudomonads in aerobic spoilage.

The possibility of retarding microbial growth on meats in defined atmospheres has also been examined.[33,49,137,165,170,188]

SIGNIFICANCE OF PROTEOLYSIS IN AEROBIC SPOILAGE

The ability of many organisms associated with spoiling flesh to demonstrate proteolytic ability on laboratory media has led to the assumption that spoilage bacteria proceed immediately to attack meat proteins with the production of peptides and amino acids which are then further degraded.

Whether or not this ability is a significant factor in spoilage is a question which has been the subject of many conflicting research reports.

The first evidence that proteolysis was not involved in the production of spoilage odours was provided by Lerke et al.[110] These workers suggested that low molecular weight nitrogenous compounds were the substrates degraded to produce the odoriferous compounds associated with spoilage of fish muscle. This evidence was subsequently supported for red meat[37,88,94] and for poultry meat.[3] However, other reports[19,20,42,73,74,181,182] have demonstrated proteolysis of both myofibrillar and sarcoplasmic proteins by various bacterial species and suggested that proteolysis may be involved in spoilage processes. Indeed Sikes and Maxcy[178] were firmly of the opinion of the importance of proteolysis and developed a system for the differentiation of food spoilage bacteria on the basis of their ability to utilise different proteins.

There is no doubt that many spoilage bacteria elaborate extracellular proteases which can degrade both sarcoplasmic and myofibrillar proteins, however, the argument is not whether proteolysis does or does not occur, but at what time and what is the contribution to off odour?

The experiments in which proteolysis has been demonstrated have used a variety of techniques, particularly starch and disc gel electrophoresis,[19,20,73,74,181] while Dutson et al.[42] used transmission electron microscopy to examine the effect of *P. fragi* on myofibrillar protein.

Dainty et al.[37] criticised these experiments on the basis that spoilage odours were apparent at the earliest sampling times used and no data relating to the prespoilage period were reported. Thus in the experiments of Hasegawa et al.[73,74] and Dutson et al.[42] bacterial numbers at the first sampling time were $> 10^9/\text{g}$ whereas Tarrant et al.[181] reported organoleptic spoilage on day 5 at 10°C when numbers were also $c. 10^9/\text{g}$, but did not examine protein breakdown until after 8 days incubation at $10\,^{\circ}\text{C}$.

Dainty et al.[37] therefore employed sampling procedures to overcome this difficulty, used muscle slices instead of muscle minces and compared the effects of the natural spoilage flora, an inoculum of beef slime and several pure cultures. Protein breakdown was not observed in either slime inoculated or naturally contaminated beef until after spoilage and bacterial slime were observed. However, extensive post spoilage proteolysis due to bacterial activity was observed. Neither the slime inoculum nor pure cultures produced a natural spoilage pattern and the authors suggest that conclusions drawn from experiments with other types of inocula cannot be taken to represent natural spoilage. The data provided by Dainty et al.[37] provides the best proof to date that while proteolysis does occur during natural spoilage it plays no significant part in the production of off odours on meat held aerobically at chill temperatures. The experimental conditions using an intact substrate with the natural spoilage flora provided a spoilage microenvironment as close to the practical situation as possible.

The results support the contentions of Lerke et al.[110] who used fractionated fish muscle press juice as a substrate and found that while the non-protein fraction (mol. wt < 5000) spoiled according to the normal organoleptic and chemical indices the protein fraction remained bland and unspoiled throughout the storage period. Whilst the criticisms made by Herbert et al.[77] about the use of artificial substrates may be applicable to the experiments of Lerke et al.[110] this remained the least equivocal evidence for the role of protein in spoilage until the work of Dainty et al.[37] It seems strange that in the interim their experimental design was not used as the basis for work on other meat substrates.

Although the experimental evidence leans heavily in favour of the lack of significant proteolysis of myofibrillar or sarcoplasmic protein before spoilage, Sikes and Maxcy[178] still suggest that 'highly proteolytic spoilage bacteria pose the greatest threat to the storage stability of refrigerated meats'. However, their interpretation of data from the literature in favour of proteolysis seems rather dubious. This is based on results [19,20,73,74,182] which have been criticised by Dainty et al.[37] Indeed while Dainty et al. have been cited as showing that under natural spoilage conditions only tropomyosin of the myofibrillar proteins was degraded, no mention is made of the major conclusions from their work. Also, the method developed for the differentiation of spoilage bacteria employs proteins as the sole source of carbon and nitrogen, a situation rather different from that found in a natural meat substrate where an adequate supply of low molecular weight substances is to be found.[106] While Sikes and Maxcy[178] suggest that as the easily assimilable nutrients in the system decrease, protein metabolism becomes essential to the propagation of those members of flora possessing proteases, this is a post spoilage phenomenon.[36]

A different type of experiment on the occurrence and significance of proteolysis was carried out by Gill and Penney[64] who considered the extent to which bacteria might penetrate meat. This was thought to be important both for reasons of hygiene and understanding spoilage processes.

While spoilage at chill temperatures is generally considered to be a surface phenomenon some workers have suggested that bacteria may penetrate 10–15 cm into meat.[47] Others, however, could not detect bacteria at depths > 2 cm in normal meat but these were present throughout meat from animals injected with papain before slaughter.[185]

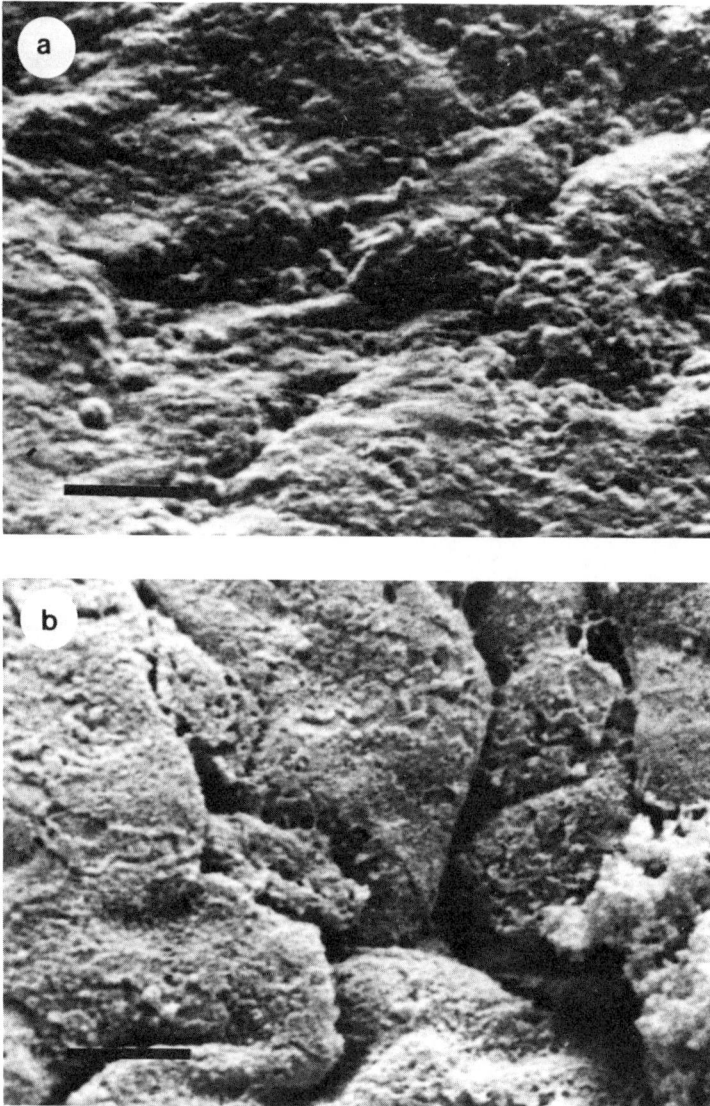

FIG. 2. Scanning electron micrograph of broiler carcass skin immediately after processing. (a) The entire surface is covered with a liquid film which has been preserved *in situ* by fixation in osmium tetroxide vapour. (b) Liquid film removed, tissue fixed in glutaraldehyde. Bars = 5 μm.

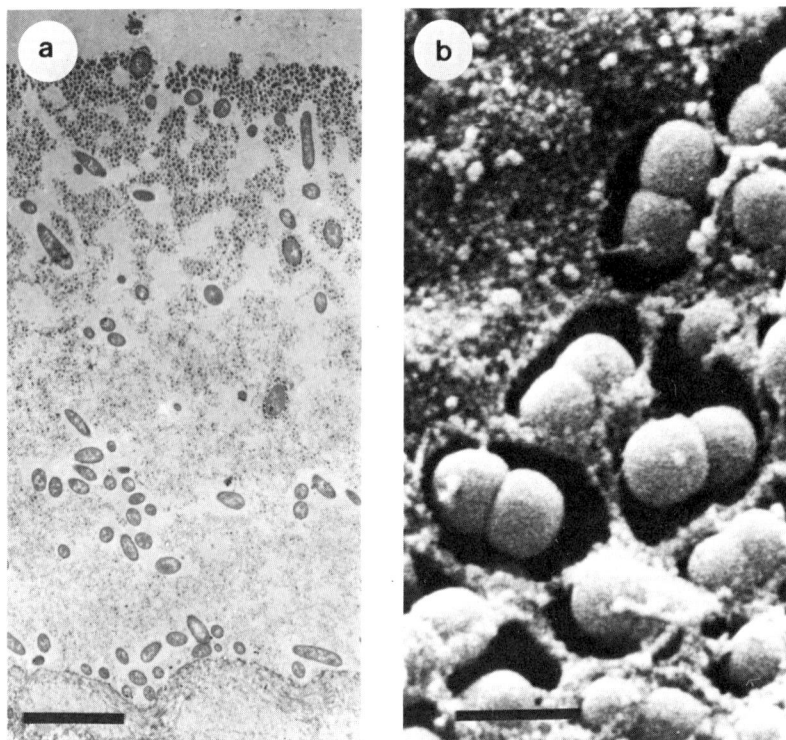

FIG. 3. (a) Transmission electron micrograph through broiler carcass skin after 16 days storage at 2 °C to show bacteria developing within the preserved liquid film on dermal skin tissue. Skin fixed in glutaraldehyde containing alcian blue and post fixed in osmium tetroxide. Sections stained with uranyl acetate/lead citrate. Bar = 5 μm. (Micrograph courtesy of Dr C. J. Thomas.) (b) Scanning electron micrograph of bacteria on broiler carcass skin after 5 days storage at 5 °C to show localised shrinkage of liquid film components. Specimen fixed in osmium tetroxide vapour. Bar = 4 μm.

An intriguing feature of Gill and Penney's work was that light micrographs were provided to supplement the bacteriological data. These demonstrated that penetration resulted from breakdown of connective tissues between muscle fibres by proteolytic enzymes secreted by the bacteria. Non-proteolytic organisms, however, were unable to invade even when grown in association with the proteolytic types. The authors concluded that no penetration should occur in organoleptically sound meat as proteases are not produced until the end of the logarithmic phase of growth. Unfortunately the incubation temperatures selected for some strains did not include one in the chill range and it may not be valid to extrapolate the results to spoilage processes at refrigeration temperatures.

Shewan[174] noted the lack of knowledge on how bacteria penetrate through fish skin and indeed on the structure and histology of fish skin. Subsequent studies at the Torry Research Station suggested that penetration occurred at temperatures above 8 °C but was much more restricted at lower incubation temperatures.[175]

Thomas[184] carried out the first major study of histological/micro-biological relationships of flesh foods and this has provided novel information on both the contamination of broiler carcass skin during processing and the subsequent development of the spoilage flora on that substrate. In fact this study has raised some fundamental questions which await elucidation and which call into question the results of earlier investigators. Central to spoilage studies is the development of the spoilage association in a fluid layer retained on the skin surface after processing (Figs 2(a), (b)). It is from the constituents of this layer that the spoilage flora derive their nutrition and produce the metabolic end-products associated with spoilage off odours. Therefore, earlier investigators[2] who examined the degradation of components of chicken skin *per se* may well have been investigating the wrong substrate (Figs 3(a), (b)).

The fluid layer has been shown by electrophoretic and chromatographic studies to contain proteins (particularly serum albumin) and amino acids. In effect, the layer provides an excellent growth medium and degradation of the actual skin was only rarely observed and then only after 16 days at 2 °C by which time the carcass had spoiled.[184]

PREFERENTIAL UTILISATION OF SMALL MOLECULAR WEIGHT COMPOUNDS

While low molecular weight substances are utilised in preference to proteins, it appears that the spoilage flora also show preferential utilisation of substrates within the low molecular weight group.

Gill[60] examined the preferential utilisation of substrates in meat juice medium and on meat by a fluorescent *Pseudomonas* sp. (aerobe) and a *Lactobacillus* sp. (anaerobe). Growth of the *Lactobacillus* sp. in meat juice medium ceased when glucose was exhausted and growth on the meat surface was limited by the rate of diffusion of fermentable substrates from within the meat. The aerobe, however, utilised amino acids and lactic acid when the glucose disappeared at c. 3×10^8 cells/cm^2. Thereafter pH and NH$_3$ concentration increased due to the deamination of various amino acids which led to the production of off odours and flavours.

Gill and Newton[61] have also examined the preferences of the aerobic spoilage flora of meat at chill temperatures. Glucose was utilised preferentially by all species except *Acinetobacter* which preferred amino acids and was inhibited by a pH of <5.7. After glucose depletion *B. thermosphacta* utilised glutamate, *Enterobacter* utilised glucose-6-phosphate and *Pseudomonas* utilised amino acids and lactic acid. None of the bacteria ceased growth because of substrate exhaustion at the meat surface and oxygen availability was suggested to be the limiting factor. During growth of mixed cultures it was demonstrated that there were no interactions until one organism had attained its maximum cell density. The pseudomonads predominated because of their faster growth rates and exhibited generation times c. 30% less than the other genera.

Under anaerobic conditions Newton and Gill[134] found that none of the bacteria studied utilised more than two of the low molecular components of meat. *B. thermosphacta* metabolised only glucose, *Enterobacter* used glucose and glucose-6-phosphate and *Lactobacillus* utilised glucose and arginine. Affinity for the common substrate, glucose, occurred in the sequence *Enterobacter* > *B. thermosphacta* > *Lactobacillus*. This led to inhibition of *B. thermosphacta* by *Enterobacter* under glucose limiting conditions but *Lactobacillus* overcame its deficiency in affinity for glucose by its ability to utilise arginine and by producing an antimicrobial substance which inhibited competitors. Thus on meat stored anaerobically *Lactobacillus* was predominant at all temperatures from 2° to 15 °C.

The preferential utilisation of some low molecular weight compounds, particularly glucose, has important implications for the keeping quality of meat and may be used to advantage to extend the shelf life of the product.

Shelef[171] in an examination of the spoilage of fresh beef liver, recorded spoilage after 7–10 days at 5 °C at population levels of c. 7–8 $\times 10^7$/g. Spoilage was due to souring, with lactic acid bacteria predominant and Gram-negative organisms c. 10^6/g. The result was explained by the carbohydrate content of the liver (c. 5% w/w), and similar spoilage patterns have previously been noted for other flesh foods containing high levels of carbohydrate e.g. oysters.[81,155]

Subsequently, Shelef[172] considered the effect of added glucose on the spoilage characteristics of ground beef. Samples with <2% added glucose spoiled typically with Gram-negative psychrotrophs predominant. However, in samples containing 2–10% glucose, the production of acid lowered the meat pH from 5·8 to 5·0–5·2 which inhibited the growth of pseudomonads. After the glucose had been utilised the pH started to rise and normal spoilage ensued. Flora analyses indicated that lactobacilli counts were only slightly higher in glucose treated meats and that the added glucose had caused a shift in nutrient utilisation by the spoilage flora. The pseudomonads metabolised the carbohydrate preferentially with the formation of organic acids resulting in a lowered pH. In practical terms, addition of 2% glucose to ground beef increased shelf life from 5 days to 8–10 days at 5°C.

Conversely, Newton and Gill[135] have reported that dark, firm, dry meat which contains little or no glucose and has a pH c. 6·0 spoils more readily than meat of normal ultimate pH (pH 5·5). Previously Gill and Newton[61] had demonstrated that the growth rates of most meat spoilage bacteria (except *Acinetobacter*) were unaffected by pH between 5·5 and 7·0. Similarly, Barnes and Impey[11] reported that fluorescent and non-fluorescent pseudomonads developed equally well at pH values between 5·8–7·0 but that the generation time of *Alteromonas putrefaciens* was doubled at pH 5·8 compared with pH 7·0 and that *Acinetobacter* would not grow at pH 5·8. Newton and Gill[135] therefore examined the hypothesis that in meat devoid of glucose the spoilage flora proceed immediately to degrade amino acids to satisfy their energetic requirements with concomitantly earlier production of NH_3 and other odoriferous by-products. By adjusting the pH and glucose content of muscle samples they were able to show that glucose concentration and not pH was the critical factor for keeping quality. Off odours could be detected in the meat samples lacking glucose after 2 days but were detected after 4 days in samples containing glucose.

Meat derived from the carcasses of stressed animals with a high ultimate pH also has important implications for vacuum packaged meats. Vacuum packaged meat of normal pH stored at chill temperatures supports the growth of a flora predominantly of *Lactobacillus* species. However, vacuum packaged high pH meat will also allow the development of some Gram-negative groups more often associated with aerobically stored material.

The spoilage condition known as greening was first reported in vacuum packaged high pH meat by Nicol *et al.*[139] and was shown to be due to the reaction of sulphides, produced by *P. mephitica* (syn. *A. putrefaciens*), with meat pigments. The condition developed only in meat with a pH > 6·0 at reduced oxygen tensions. However, work on preferential utilisation of nutrients and the extension of shelf life of aerobic high pH meat by addition of glucose, but not reduction of pH[60,61,134,135] raised the question of which was the more important parameter in the control of spoilage of vacuum packaged high pH meat. Subsequently, Gill and Newton[63] concluded that production of spoilage odours by *E. liquefaciens* could be prevented by addition of glucose, but that greening of meat by *A. putrefaciens* was not prevented as the organism degraded cysteine with the release of H_2S even when glucose was present. Prevention of greening required inhibition of *A. putrefaciens* by pH reduction to below 6·0.

The studies of Gill and coworkers at the New Zealand Meat Research Institute represent a marked advance in our understanding of the mechanisms of meat spoilage. Further studies of this nature are required on other substrates, e.g. the layer material on chicken skin described by Thomas.[184] Detailed examinations of the degradation of this material or other 'meat juice' substrates in combination with sophisticated conductance measurements[158] capable of showing step-wise development of the population on preferentially utilised substrates offer the possibility of further advances in the determination of the precise mechanisms of spoilage.

STERILE SUBSTRATE STUDIES

Much of the work on off odour production now relies on the excision of sterile muscle sections as spoilage substrates. The reasons for the use of a natural substrate have been listed by Herbert *et al.*[77] Thus it is possible to determine the potential contribution of individual strains and mixtures of strains to the spoilage aroma and relate this to the volatiles produced during natural mixed culture spoilage. Dainty *et al.*[37] presented a logical argument to suggest that natural spoilage produced a unique biochemical degradation pattern in relation to protein breakdown which could not be reproduced by individual cultures or 'slime' inoculum. No doubt the same type of philosophical argument might be applied to the study of spoilage volatiles. However, in this extremely complex situation it is necessary to simplify the equation either in terms of the substrate or the inoculum in order to build up data which can be related to natural spoilage.

Difficulty arises when it is not possible to obtain sterile samples of the substrate, e.g. hamburger meat,[117] chicken skin.[41] In the latter case electron microscopic observations of the substrate have shown that the organisms are located in a layer of material on the skin surface and may be located in spaces of capillary size.[119,184] For these reasons sterile skin could not be consistently obtained and the spoilage potential of skin isolates was assessed on muscle sections. However, Thomas[184] provided an alternative natural substrate, psychrotroph free skin, which was obtained by a scalding treatment similar to that used in commercial practice (60 °C, 3 min). Thus if the skin samples were stored at chill temperatures until use and inoculated with a psychrotrophic strain only this could develop and produce off odours.

The use of sterile muscle sections has also allowed characterisation of the types and proportion of off odour producers at various stages of spoilage. Once again the experimental design for work on poultry substrates[40,114,115] has been developed from examinations of fish spoilage[76] and data for red meat is noticeably lacking. Work with fish substrates[4,77] has demonstrated that off odours were produced by a consistently small fraction of the population present at any stage of spoilage. Similarly, McMeekin[115] and Daud et al.[40] noted a similar trend for chicken leg muscle and skin. On breast muscle, McMeekin[114] recorded a selection for spoilers as storage progressed, but this was subsequently explained by the fact that the isolation procedure (colonies picked from nutrient agar after 7 days at 2 °C) recovered only a few psychrotrophic types, some of which were off odour producers and exhibited the fastest growth rates.

VOLATILE SULPHIDES IN SPOILAGE

Occurrence and Properties of Volatile Sulphides
Sulphide-like off odours are important components of the spoilage aroma of meats and have been reported in a wide variety of products including red meat,[117,139] uneviscerated poultry,[14] eviscerated poultry,[40,53,107,114,115,117] fish,[25,29,77-80,126] shellfish[103] and vacuum packaged bacon.[57] Sulphydryl compounds may also contribute substantially to the characteristic flavour and aroma of many foods; cooked brassicas,[39] cheese,[105,120-122] molluscan shellfish,[21,159] cooked poultry,[127,128] and cooked beef.[151]

An important property of these compounds is that they are detected organoleptically at extremely low concentrations. The odour thresholds and characteristics of the three principal volatile sulphides (hydrogen sulphide, H_2S, methanethiol, CH_3SH, and dimethylsulphide, $(CH_3)_2S$, have been compiled by Herbert[76] and are shown in Table 1. Dimethyl-disulphide is often also found in the headspace above spoiling flesh foods.

Mechanism of Production of Volatile Sulphides
The mechanism of production of volatile sulphides has been reviewed by

TABLE 1

ORGANOLEPTIC DESCRIPTIONS OF VARIOUS CONCENTRATIONS OF VOLATILE SULPHIDES

Compound	Concentration in aqueous solution (ppb)	Odour description	Concentration in naturally spoiling fish muscle after 10 days at 0°C (ppb)
H_2S	40	Slight H_2S *Threshold value*	150
	80	Medium to strong H_2S	
$(CH_3)_2S$	0·50	Slightly sour trace of sulphide *Threshold value*	20
	0·75	Slight cabbage water	
	1·50	Strong, sulphidy, byre like	
CH_3SH	0·05	Very slightly sour, *Threshold value*	120
	0·50	Slight cabbage water, leeks	
	2·00	Sharp, strong, stale cabbage water	
	100·00	Metallic, cooked meat, sulphidy	

Data of Herbert.[76]

Kadota and Ishida[97] and Herbert *et al.*[78] Herbert and Shewan[79,80] have provided definitive evidence for the precursors and mechanism of formation of sulphides in spoiling cod muscle. There are also reports (cited above) which indicate the production of sulphides during the spoilage of poultry and red meat and it is probably reasonable to expect similar precursors and mechanisms of production. However, detailed experimental evidence on the genesis of sulphides in these substrates is lacking.

Hydrogen sulphide is produced from cyst(e)ine as follows[31]

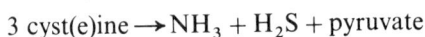

$$3 \text{ cyst(e)ine} \rightarrow NH_3 + H_2S + \text{pyruvate}$$

However, Jocelyn[96] noted that the true reactant is cystine but this is required only in catalytic amounts as the hydrosulphide intermediate oxidises more cysteine to cystine during the reaction.

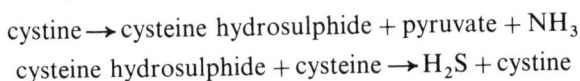

$$\text{cystine} \rightarrow \text{cysteine hydrosulphide} + \text{pyruvate} + NH_3$$
$$\text{cysteine hydrosulphide} + \text{cysteine} \rightarrow H_2S + \text{cystine}$$

$$HOOCCH(NH_2)CH_2CH_2SCH_3$$
Methionine

$$\downarrow$$

oxidative deamination

$$\downarrow$$

$$HOOCCOCH_2CH_2SCH_3 + NH_3$$
demethiolation

$$\downarrow$$

$$HOOCCOCH_2CH_3 \quad + \quad CH_3SH$$
α-ketobutyric acid   methanethiol

$$\downarrow \qquad\qquad \downarrow$$

$$CO_2 + H_2O \qquad \tfrac{1}{2}(CH_3SSCH_3)$$
dimethyldisulphide

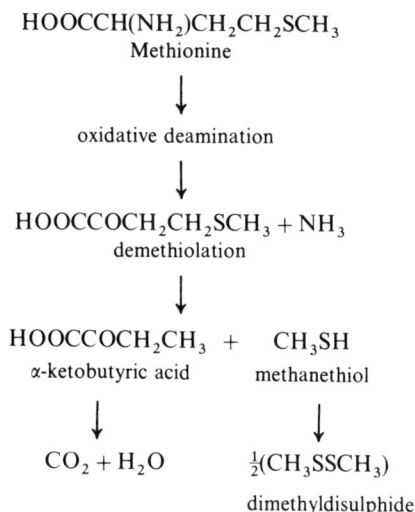

FIG. 4.   Degradation of methionine.[166]

Decomposition of methionine to methanethiol was reported by Segal and Starkey.[166] Methionine is first deaminated and the product demethiolated (Fig. 4) to form methanethiol which is subject to autooxidation or enzymatic oxidation to dimethyldisulphide.

The mechanism of formation of dimethylsulphide in spoiling flesh foods is less clear. In marine products the precursor of dimethylsulphide is often dimethyl-β-propiothetin which occurs widely in marine algae.[1] Fish or shelfish feeding on such algae may display high levels of dimethylsulphide which is formed from DMPT as follows

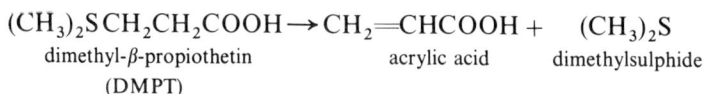

$$(CH_3)_2SCH_2CH_2COOH \rightarrow CH_2{=}CHCOOH + \quad (CH_3)_2S$$
dimethyl-β-propiothetin   acrylic acid   dimethylsulphide
(DMPT)

This may cause a spoilage condition e.g. 'blackberry' problem in cod from the Labrador area[179] or 'petroleum odour' in canned chum salmon[132] or in the cases of some molluscan shellfish be the major flavour volatile.[21,159] One may even speculate that the production of dimethylsulphide might severely limit the commercial exploitation of Antarctic Krill (*Euphasia superba*) which feed on DMPT containing algae.[177]

In spoiling meats methionine may be the precursor of dimethylsulphide as Kallio and Larson[98] have reported this degradative pathway in cell free extracts of *Pseudomonas* spp. and Challenger[30] noted the reaction with the fungus *Scopulariopsis brevicaulis*. Another possibility is the methylation of CH_3SH,[97] but present information on the enzymology of dimethylsulphide formation is unsatisfactory.

An interesting situation occurs with the sulphur containing amino acid taurine. Although this is the major free amino acid in chicken muscle[107] and in certain fish species[79] it appears to almost totally resist bacterial degradation, at least aerobically at chill temperatures. It would, therefore, be interesting to determine which organisms are capable of utilising taurine and under what conditions this could occur.

Microorganisms Involved in Sulphide Production in Chill Stored Meats
Several groups of organisms appear to be particularly associated with the production of sulphide-like off odours in spoiling flesh foods and prominent among these is the organism now referred to as *Alteromonas putrefaciens*.[108]

This organism was originally isolated from putrid butter[112] but has subsequently been reported in many other spoilage situations. Barnes and Thornley[15] in a comparison of the spoilage flora of eviscerated chickens stored at different temperatures reported *A. putrefaciens* as 19% of the flora of carcasses allowed to spoil at 1 °C with 71% fluorescent and non-fluorescent pseudomonads. Barnes and Impey[11] in an examination of the growth of *A. putrefaciens* and other psychrotrophic spoilage organisms noted sensitivity to low pH as previously described by Long and Hammer.[112] In chicken leg muscle (pH 6·65) or in broth (pH 6·2) rapid growth was recorded but the generation time was doubled to 16·5 h at 1 °C by lowering the pH to 5·8. On breast muscle (pH 5·8) mean generation times between 12 and 13·4 h were recorded. The slower rate of growth meant that *A. putrefaciens* was unable to compete with *Pseudomonas* group I and II types and played no part in the spoilage of breast muscle. On leg muscle *A. putrefaciens* developed at the same rate as the group II types and accounted for c. 20% of the flora at spoilage. It was noted that if methods are used which delayed the growth of *Pseudomonas* strains such as high levels of CO_2 then the carcass might be expected to spoil as a result of the growth of *A. putrefaciens*.

A similar pH effect was reported by McMeekin[114,115] in separate studies of the spoilage of chicken breast and leg muscle. *A putrefaciens* was absent from the spoilage association of breast muscle but was recovered as a consistently small fraction of the flora of spoiling leg muscle. All isolates were psychrotrophic and produced strong off odours when grown on chicken leg muscle.

Freeman *et al.*[53] also implicated *A. putrefaciens* in the production of compounds found in the aroma of spoiled chicken. The volatile compounds associated with spoiled chicken breasts stored for 12 days at 2 °C or 5 days at 10 °C are listed in Table 2. Two named isolates of *A. putrefaciens* produced H_2S, CH_3SH and $(CH_3)_2S_2$ but only one of three isolates caused spoilage when grown on sterile chicken breast muscle.

TABLE 2

VOLATILE COMPOUNDS ASSOCIATED WITH
SPOILED CHICKEN BREASTS STORED AT 2° OR
10 °C

Hydrogen sulphide	n-heptane
Methyl mercaptan	1-heptene
Dimethylsulphide	Heptadiene
Dimethyldisulphide	n-octane
Methyl acetate	acetone
Ethyl acetate	benzene
Methanol	toluene
Ethanol	

Data of Freeman *et al.*[53]

In a survey of the types of aerobic organisms able to produce H_2S on peptone iron agar[111] McMeekin and Patterson[117] recovered 26 strains of *A. putrefaciens* from refrigerated steak. All of the isolates grew well at 5 °C and much more quickly at this temperature than the other H_2S producing types isolated (*Proteus* sp., *Citrobacter freundii* and coryneforms). Five representative strains caused a typically sulphide-like spoilage odour when grown on sterile chicken muscle, off odours being detected organoleptically after 7 days storage at 5 °C.

Several authors have also examined the phenomenon of greening in vacuum packaged meat.[16,63,139,183] The effect of meat pH and substrate utilisation have been discussed above. However, the possible role of volatile sulphides other than hydrogen sulphide in this defect has not been considered.

Gas chromatographic studies[77–80] have demonstrated that *A. putrefaciens* produces H_2S, CH_3SH and $(CH_3)_2S_2$. Attention is drawn to this point by the continuous culture experiments of Gill and Newton.[63] These workers grew *A. putrefaciens* in a simple salts–glucose medium and detected H_2S production on lead acetate papers above culture samples supplemented with cysteine and methionine. Anaerobic and oxygen limited cultures growing at the submaximal rate evolved H_2S without delay when

TABLE 3

A COMPARISON OF THE SPOILAGE FLORA OF CHICKENS STORED AT 1 °C WRAPPED IN OXYGEN PERMEABLE FILM (A) AND OXYGEN IMPERMEABLE FILM (B)

Treatment of carcasses and storage time (days)	Colony count at 1 °C (organisms/ cm²)	Total no. of strains	% Distribution of strains				
			Pseudomonas (pigmented)	Pseudomonas (non-pigmented)	A. putrefaciens	Acinetobacter	Unclassified
Film A^a							
0	$2·7 \times 10^4$	22	18	58	19	5	
8	$4·9 \times 10^6$	21	5	92	1	3	
12	$2·8 \times 10^8$	28	7	82	3		8
Film B^b							
0	$2·7 \times 10^4$	22	5	63	9	19	4
8	$5·2 \times 10^5$	23	12	31	42	3	12
12	$6·6 \times 10^6$	26		3	69		28^c
14	$7·5 \times 10^6$	27		4	60		36
16	$5·8 \times 10^7$	26		38	51		11

[a] Oxygen permeable film (polyethylene).
[b] Oxygen impermeable film (vinylidene chloride–vinyl chloride copolymer).
[c] Cocci, possibly unidentified streptococci.
Data of Barnes and Melton.[13]

cysteine was added but H_2S was not detected, even after prolonged incubation, with added methionine. This is not unexpected considering the findings of Herbert and Shewan[79,80] and the limitations of lead acetate as a reagent for the detection of volatile sulphides.[116] It would, therefore, be of interest to examine the role of other volatile sulphides, particularly methanethiol, in the greening defect of vacuum packaged meat.

Alteromonas putrefaciens also appears to be favoured in competition with *Pseudomonas* species if chicken carcasses are packed in impermeable films.[11] A flora analysis based on the packaging experiments of Shrimpton and Barnes[176] was reported by Barnes and Melton[13] (Table 3). This indicated that carcasses wrapped in polyethylene films (permeable) spoiled due to the growth of fluorescent and non-fluorescent pseudomonads. The flora of carcasses stored in the impermeable film (vinylidene chloride–vinyl chloride copolymer) was dominated by *A. putrefaciens* (Table 3). In these packs the CO_2 concentration increased to 9–10% and the growth of the spoilage flora was significantly slower. However, as the pH of the rinse water of fresh chicken skin is *c*. 6·2–6·3[54] and Thomas[184] has demonstrated the development of the spoilage association in the liquid film on the skin, growth of *A. putrefaciens* may be expected on vacuum packaged, chill stored chicken carcasses.

Fluorescent pseudomonads have also been reported to produce sulphides on chill stored flesh foods. Again the first reports were for fish substrates,[77] but subsequently these organisms were shown to produce sulphides from chicken breast muscle,[53,114] chicken leg muscle[115] and chicken skin.[40,184]

The latter study also reported *Pseudomonas* group II (non-fluorescent) isolates to produce sulphide-like off odours when grown on sterile chicken tissues as previously reported by McMeekin.[114,115]

Other organisms which produce volatile sulphides may also be recovered from spoiling meats. These include psychrotrophic enterics such as *Proteus, Citrobacter, Hafnia* and *Serratia*[44,75,117,152–154] and flavobacteria.[53,113] Although the pseudomonads are of particular importance at chill temperatures, the other types might well contribute to sulphide-like spoilage odours if ecological conditions are imposed which limit the growth of the pseudomonads or the storage temperature rises out of the chill range. Patterson and Gibbs[152,153] refer particularly to the development of psychrotrophic enterics at storage temperatures of 7°C and the effect of 'conditioning' treatments designed to produce tender meat. Similarly conditions for the proliferation of these types may occur during hot boning procedures.

An interesting spoilage condition of vacuum packaged bacon involving *Proteus inconstans* was reported by Gardner and Patterson.[57] This organism caused a spoilage aroma referred to as 'cabbage odour' which was due to methanethiol production. Inoculation of various vacuum packed bacons, with *P. inconstans*, suggested that methanethiol was more likely to be produced in bacon of pH $>6\cdot0$ and salt content $<4\%$. The authors point out that such a salt content is specified for mild cured bacon and that a pH $>6\cdot0$ in cured meats is common. Storage time and temperature were regarded as the important control parameters as methanethiol was detectable in inoculated packs after 4 days at 20 °C but not after 4 weeks at 4 °C.

McMeekin and Patterson[117] isolated a number of hydrogen sulphide producing coryneform bacteria from hamburger meat but these were not considered important spoilers at chill temperatures. However, it is of interest that Sharpe *et al.*[167,168] have isolated coryneforms from human and dairy sources which produce methanethiol and Law and Sharpe[105] have considered their role in cheese flavour.

Methods for the detection and isolation of sulphide producing microorganisms in foods have traditionally relied upon the use of lead or iron salts.[34,40,77,111,117] However, these are of limited value in that they detect only hydrogen sulphide producing organisms and those which produce sulphides other than hydrogen sulphide would not be detected by metal salts. This may be of considerable importance as Thomas[184] demonstrated that methanethiol and not H_2S was the predominant sulphide produced by organisms causing sulphide-like off odours on chicken muscle. Twenty-eight such isolates produced methanethiol but only three of these gave lead acetate positive reactions.

An alternative is provided by the use of Ellman's reagent, 5,5′dithio-bis-2-nitrobenzoic acid (DTNB) which reacts with aliphatic thiols to yield a highly coloured (yellow) aromatic thiol (Fig. 5).[46] The reagent will not detect dimethylsulphide or dimethyldisulphide but this is not a serious limitation as these compounds are normally found as minor components in association with methanethiol.

Use of the reagent has been reported in three studies involving food systems[65,116,168] and in each a different method was employed. Sharpe *et al.*[168] used the method of Laakso[101] in which washed cell suspensions were incubated with methionine and DTNB. Gillespie[65] described a procedure in which colonies grown on a nutrient agar were covered with a film of plain, unlaquered cellophane on which a filter paper soaked in DTNB was placed. Sulphide producers were identified by the formation of a yellow colouration

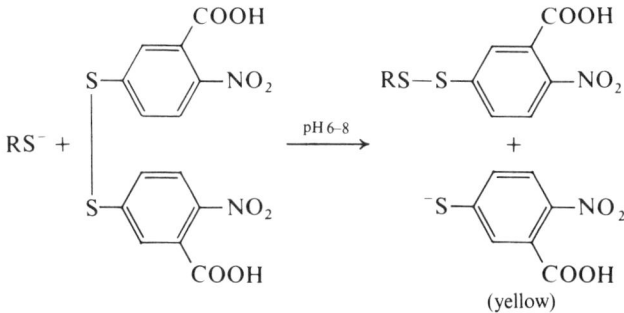

FIG. 5.   Reaction of DTNB with sulphides.[46]

on the filter pad adjacent to the colony. Unfortunately the method does not easily allow recovery of organisms of interest from the primary isolation plates and in this laboratory repeated attempts have been made to develop a direct isolation and enumeration procedure based on DTNB, e.g. using the Anderson and Baird-Parker[5] technique of growing the colonies on cellulose acetate membrane filters which were then placed on a reagent soaked filter pad. However, all have been unsuccessful because of false positive reactions. There appear to be three problems:

(1) Protein digests and hydrolysates used in complex media contain cysteine which reacts with the reagent.
(2) Reactions may occur with sulphydryl groups closely associated with the cell.
(3) Colonies growing on complex media raise the pH above 8·0 which causes dissociation of DTNB to yield the yellow coloured anion.

Because of these constraints on a direct plating procedure, McMeekin et al.[116] preferred to detect sulphide production in the headspace above a liquid medium supplemented with methionine and/or cyst(e)ine. This was achieved by the use of DTNB impregnated paper strips and the volatile sulphide products were, in part, distinguished by the additional use of lead acetate papers and confirmed by the use of specific substrates.

OTHER OFF ODOURS

In addition to the sulphides, several other off odours have been associated with spoiling flesh foods. Prominent among these are the fruity odours normally produced by *Pseudomonas* group II types similar to *P. fragi*. As with the sulphides, the most detailed information has come from work on fish.[26,77,125] The compounds responsible for the fruity odour were reported to be ethyl esters of acetic, butyric and hexanoic acids and *P. fragi* also produced other volatile compounds, including methanethiol, when grown on sterile fish muscle.

Reports for poultry meat are mostly limited to the subjective description of fruity off odours.[10,114,115] However, Freeman *et al.*[53] provided GLC–mass spectrometry data which showed the production of esters by various organisms growing axenically on sterile chicken breast muscle. Ethyl acetate was produced by *P. fluorescens* and a *Moraxella* sp. but only the *P. fragi* cultures produced ethanol, ethyl acetate, methanol and methyl acetate.

While many of the volatiles which contribute to spoilage odours (NH_3, H_2S, CH_3SH) arise as a result of deamination of amino acids, members of the *Enterobacteriaceae* also possess the ability to decarboxylate amino acids.[35] Odoriferous compounds may also be produced by this mode of degradation of amino acids, e.g. cadaverine from lysine, putrescine from ornithine. It is also of interest that *A. putrefaciens* possesses ornithine decarboxylase[108] unlike the pseudomonads which degrade amino acids by deamination. Therefore, it may be worthwhile to examine the role of such compounds in the spoilage of meats in which enterics and *A. putrefaciens* are involved. In this context it is of interest to note that Dainty *et al.*[38] reported high levels of trimethylamine in vacuum packaged high pH meat.

Other subjective descriptions of off odours include evaporated milk[40,114] produced mainly by *Pseudomonas* group II types and a fishy odour produced by *Acinetobacter*.[10,115] These odours have not been further characterised and the need for further studies as reported by Miller *et al.*[124–126] for fish and Freeman *et al.*[53] for poultry is obvious. Dainty *et al.*[38] have obtained evidence that the sour/cheesy odours of normal pH vacuum packaged beef may in part be attributed to short chain fatty acids. Pure culture studies suggested that these may be produced by lactic acid bacteria and possibly *Brocothrix thermosphacta*. However, apart from the studies of Gibbs *et al.*[59] and Dainty and Hibbard[36] the lack of detailed data on volatile compounds produced during the aerobic spoilage of red meat represents a serious deficiency in our knowledge of spoilage processes.

MICROBIAL INTERACTIONS

Although microbial interactions and the specific growth rate of an organism were listed by Mossel[131] as implicit factors controlling the development of the spoilage association of foods, there is little information available about microbial interactions during the spoilage of meats. However, Gill and coworkers did consider the possibility of competition during aerobic and anaerobic chill storage and Gill and Newton[61] reported that *Pseudomonas* spp. dominated under aerobic conditions because of their more rapid growth rates. No interactions between species were observed until the maximum cell density was reached when *Pseudomonas* reduced the growth rate and cell crop of competitors which were unable to compete effectively for oxygen. Under anaerobic conditions *Lactobacillus* grew fastest at all temperatures between 2° and 15°C and became the dominant fraction of the spoilage association despite showing the lowest affinity for glucose which was the only substrate metabolised by all members of the anaerobic flora.[134] *Lactobacillus* was apparently advantaged by the production of an undefined antimicrobial agent as previously reported by Sandine *et al.*,[163] Gilliland and Speck[66] and Roth and Clark.[161]

This effect was applied by Raccach and Baker[156] and Raccach *et al.*[157] in an attempt to control spoilage (and pathogenic bacteria) in mechanically debonded poultry meat and ground breast muscle. Lactic acid starter cultures *Pediococcus cerevisiae* ('Accel') and *Lactobacillus* ('Lactacel DS') were used separately (10^9 cells/g) and in a 50:50 mixture (2×10^9 cells/g) to repress pure cultures of spoilage bacteria (*P. fluorescens*, *P. fragi*, *A. putrefaciens*) on cooked, mechanically deboned poultry meat.[156] In each case the individual cultures were less effective repressors than the 50:50 mixtures. A noticeable effect on each spoilage organism was the immediate effect on the inoculum which was reduced from $c.\ 10^{4.5}$ to $c.\ 10^{3.5}$ indicating the presence of an antibacterial agent at that time.

A similar effect was demonstrated on the development of the psychrotrophic population on uncooked, mechanically deboned poultry meat.[157] Again meat treated with the 50:50 inoculum showed a significantly extended shelf life compared with the control. However, the effect of individual cultures might well be explained only on the basis of the initial drop in inoculum potential. *P. cerevisiae* caused a drop in the psychrotrophic inoculum from $c.\ 10^5$ to $c.\ 10^{4.5}$ g with respective cell densities of $c.\ 3 \times 10^9$ and 8×10^8 cells/g after 7 days at 3°C. In each case this represents 14–15 generations. Time to reach 10^7 cells (the criterion of spoilage) was 4 days for the control and 5 days for the *P. cerevisiae* treatment. Again this represents similar generation times of 14 h and 15 h respectively.

With ground breast meat the inoculum caused a noticeable drop in the initial number of psychrotrophic contaminants in all treatments ($10^{4.5}$ to $10^{3.5}$/g). A lag phase of 4 days occurred with the control and the maximum cell density attained was about 10^8 cells/g compared with $> 10^9$/g reported by McMeekin[114] for intact muscle at 2 °C. This perhaps suggests an inherent inhibitory effect of the ground substrate which could not be explained by a lowered pH value of the meat. However, the 50:50 inoculum increased the lag phase from 4 to 7 days and decreased the final cell density to 10^7 cells/g after 12 days at 3 °C. In all of the experiments the effect of the individual lactic acid bacteria compared with the 50:50 mixture cannot be distinguished absolutely from the effect of doubling the inoculum from 10^8 to 2×10^8 or 10^9 to 2×10^9 cells/g or ml.

Microbial antagonisms take place in the form of competition, amensalism, predation and parasitism.[123] The role of the latter two interactions is well documented in water bacteriology but their significance in food microbiology is uncertain except for the role of bacteriophage in the lysis of starter cultures.

Neal and Banwart[133] examined the effectiveness of the small parasitic bacterium *Bdellovibrio* in the reduction of numbers of Gram-negative bacteria in foods. Sixty-eight food samples were tested by the double layer plate technique and plaques were observed on nineteen samples. Twelve of these were produced by myxobacteria, six by amoebic protozoa, one contained both organisms but no *Bdellovibrio* were found. *Bdellovibrio* were added to ground beef and pasteurised milk to determine if the parasite would reduce the natural flora. After 1 day at room temperature the meat was spoiled and the *Bdellovibrio* count decreased from 2×10^6/g to 3×10^4/g in 4 days. At 4 °C survival was 65 % but again there was no effect on the Gram-negative flora.

In this laboratory preliminary enrichment experiments in which *E. coli* was added to poultry plant immersion chiller water revealed the decline of *E. coli* and the development of plaque forming units (ciliate protozoa and myxobacteria). However, the possibility and advisability of using these predators and parasites in a practical situation remains to be assessed. For further information on microbial interactions in foods, readers are referred to the review of Kraft *et al.*[100]

EVALUATION OF SPOILAGE

Methods for the evaluation of spoilage of flesh foods have been reviewed by several authors.[84,91,93] The latter author listed 16 categories of chemical methods, 7 categories of physical methods, 5 categories of bacteriological methods and 4 categories of physicochemical methods which represented approximately 45 different techniques for the determination of spoilage (Table 4).

To be suitable for routine purposes in a practical situation, the property should be rapidly measurable, reproducible and not require the use of elaborate or expensive equipment. In addition, the technique should give a measure of the degree of spoilage and not simply diagnose rank spoilage which in any case could be detected organoleptically. The last criterion places a severe constraint on most of the physical and chemical measures so far proposed.

Chemical, Physical, Physicochemical and Bacteriological Indices

Many chemical indices of spoilage have been based on measurements of spoilage substrates (amino acids, nucleotides) or products of bacterial degradation (ammonia, sulphides, mercaptans). Unfortunately none of the proposed tests has any predictive value and measurable changes are detectable only at bacterial levels above those known to cause organoleptic spoilage (10^8–10^9/g).[59,94] This led Lea et al.[107] to the erroneous conclusion that autolytic changes were of the same order of importance in spoilage procedure as microbially induced changes. In addition, the measured changes are often inconsistent and components such as individual amino acids may increase or decrease during spoilage.[94,162] Adamcic et al.[2] were in fact able to manipulate amino acid contents depending on the type of inoculum used to inoculate the chicken skin substrate.

An explanation for the lack of measurable changes in the precursors of odoriferous compounds until high bacterial numbers have been reached has been provided by the work of Gill[60] who demonstrated under aerobic conditions the lack of degradation of amino acids until glucose had been consumed by which time the count was 10^8/cm^2.

This perhaps suggests that determination of the carbohydrate content of meat might provide a better indication of the degree of spoilage but Gardner[56] also found that changes in the carbohydrate component were of little value in spoilage evaluation. Gill and Newton[62] noted that most workers have been unable to observe glucose utilisation because they analysed for total carbohydrate using anthrone or similar reagents which react with many sugars and sugar derivatives. The measure of 'total carbohydrate' includes glycogen which is not utilised by the spoilage bacteria and glucose-6-phosphate which is used only by enteric types.[61]

Similarly, physical measurements of the degree of spoilage tend to be

TABLE 4
CATEGORIES OF METHODS FOR THE EVALUATION OF SPOILAGE[93]

Chemical methods
(a)   Measurement of H_2S production
(b)   Measurement of mercaptans produced
(c)   Determination of non-coagulable nitrogen
(d)   Determination of di- and trimethylamines
(e)   Determination of tyrosine complexes
(f)   Determination of indole and skatole
(g)   Determination of amino acids
(h)   Determination of volatile reducing substances
(i)   Determination of amino nitrogen
(j)   Determination of BOD
(k)   Determination of nitrate reduction
(l)   Measurement of total nitrogen
(m)   Measurement of catalase
(n)   Determination of creatinine content
(o)   Determination of dye reducing capacity
(p)   Measurement of hypoxanthine

Physical methods
(a)   Measurement of pH changes
(b)   Measurement of refractive index of muscle juices
(c)   Determination of alteration in electrical conductivity
(d)   Measurement of surface tension
(e)   Measurement of UV illumination (fluorescence)
(f)   Determination of surface charges
(g)   Determination of cryoscopic properties

Direct bacteriological methods
(a)   Determination of total aerobes
(b)   Determination of total anaerobes
(c)   Determination of ratio of total aerobes to anaerobes
(d)   Determination of one or more of above at different temperatures
(e)   Determination of Gram-negative endotoxins

Physicochemical methods
(a)   Determination of extract release volume
(b)   Determination of water holding capacity
(c)   Determination of viscosity
(d)   Determination of meat swelling capacity

diagnostic rather than predictive. However, rapid advances are being made in hardware designed to measure changes in the electrical conductivity (conductance measurements) of bacterial cultures during growth. Richards et al.[158] have described the construction of electronic equipment capable of detecting several hundred bacteria/ml with up to 128 channels and with computer plotting of results. This type of equipment provides the basis for very rapid automated analyses of samples.

The physicochemical measure which has received most attention is extract release volume (ERV). Meat of good organoleptic quality releases a large volume of extract which decreases with the onset of spoilage. Jay[88−90] and other authors have claimed a predictive value for the test.[18] Ingram and Dainty,[84] however, suggested that the ERV value of c. 30 given by Jay and Borton to discriminate between acceptable and spoiled meat was related to bacterial numbers and organoleptic characteristics associated with incipient or frank spoilage and as such the predictive value of the test was little better than other methods.

In addition to favouring the use of ERV as a spoilage indicator Jay[93] pointed out that the method revealed two aspects of the spoilage mechanism:

(1) that aerobic spoilage occurs in the absence of proteolysis as the ERV is decreased rather than increased as is the case when meat is treated with proteinases;
(2) that a marked increase in hydration capacity of meat proteins occurs during spoilage.

The mechanism by which this occurs remains obscure although Shelef and Jay[173] have implicated amino sugar complexes produced by the spoilage flora. Once more Ingram and Dainty[84] raised objections to this theory and pointed out that amino sugars are components of the peptidoglycan layer of all bacteria (except Mycoplasma and l-forms). It was suggested that the concentrations of glucosamine and other amino sugars which Shelef and Jay[173] found present as polymers and which were measured after hydrolysis simply reflected bacterial numbers and hence ERV.

Direct bacteriological methods while useful in an ongoing quality assurance situation, suffer from the obvious drawback that 48–72 h incubation is required before colony forming units can be satisfactorily enumerated. Jay[92] may have overcome this difficulty by the application of an extremely sensitive Limulus lysate test (LLT) technique which measures the presence of Gram-negative endotoxins. This is based on the use of a lysate of amoebocytes of Limulus (the horse shoe crab) which gels in the presence of Gram-negative bacterial cells and can detect pg quantities of endotoxin. The advantages claimed are speed and simplicity, the test responds only to Gram-negative bacteria which are the significant spoilage group during aerobic storage, and it correlates well with ERV measurements. A simple screening test for refrigerated ground beef may be performed by preparing a 10^{-1} dilution of the sample and testing a 10^{-3} dilution of the homogenate. A positive result indicates that the meat has spoiled or is at a stage of incipient spoilage, a negative result indicates a satisfactory sample.

Jay's[92] modification of the LLT from urine analysis to spoilage studies suggests that it may be possible to examine the application of other characteristics of the significant flora as indices of spoilage . Several reports have shown the enzyme aminopeptidase to be present in Gram-negative cells[27,28] and Gray and Miller[67] have used a colorimetric method based on a carbocyanine dye 1-ethyl-2-]3-(1-ethylnaphtho[1,2d]-thiazolin-2-ylidene)-2-methylpropenyl]naphtho[1,2d]-thiazolium bromide which measures the lipopolysaccharide fraction of Gram-negative cell walls, to enumerate bacteria in urine. These methods have not, as yet, been applied to spoilage studies but may be worth consideration.

Temperature Function Integration
In addition to the spoilage indices outlined above, several attempts have been made to predict the shelf life of aerobically stored products on the basis of temperature history.[149,150,180] This approach is based on the fact that temperature of storage is the cardinal factor controlling the rate of development of the microflora and thus the rate of spoilage of meats.

Spencer and Baines[180] postulated a linear relationship between temperature and rate of spoilage in the range -1 to $25\,^\circ$C. This was described by the formula

$$k_t = k_0(1 + c\theta)$$

where k_t is the rate of spoilage at temperature $\theta\,^\circ$C, k_0 is the rate of spoilage at $0\,^\circ$C and c is a constant for linear response. The linear response was used by Ronsivalli and Charm[160] as the basis of a shelf life prediction slide rule to measure the remaining shelf life of cod held at temperatures from 0 to $8\,^\circ$C.

An alternative hypothesis was proposed by Olley and Ratkowsky.[149] This suggested that the data fitted the Arrhenius equation in which spoilage rate was substituted for growth rate. Seventy data sets from the literature, split into 249 subsets, were used and for each set values of μ (apparent activation energy) and c (constant for linear response) were determined.

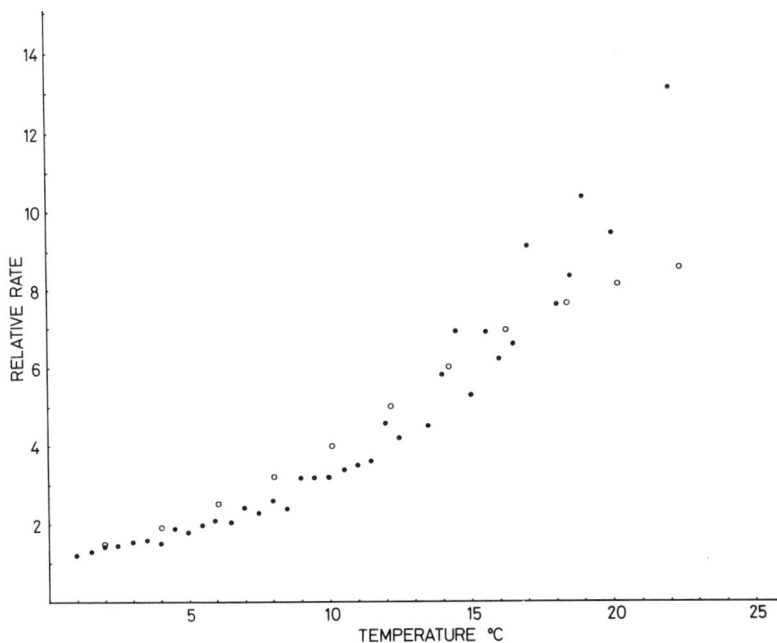

FIG. 6. Comparison of mean relative rates of growth and metabolism of poultry spoilage bacteria (●), with relative rates predicted by the general spoilage curve of Olley and Ratkowsky,[149,150] (○). Data of Daud et al.[41]

Although Spencer and Baines[80] suggested that c was constant between -1 and 25 °C, Olley and Ratkowsky[149] showed c varying from 0·24 at 1·8 °C to 0·46 at 13·5 °C. The value was constant at c. 0·24 with a small standard deviation up to 6 °C but above this temperature the value of c increased and became more variable. The apparent activation energy decreased with temperature as expected and had a much lower standard deviation at the higher temperature ranges (Table 5). A relative spoilage rate curve (Fig. 6) was constructed over the temperature range 0°–25 °C from the average activation energies ($\bar{\mu}$) and corresponding mean maximum temperature (Table 5). Relative spoilage rate was defined as the ratio of the actual spoilage at any temperature to the actual rate at 0 °C. The validity of the curve is further suggested by computation of relative rates at 2500 cal/mol intervals of $\bar{\mu}$ which indicates that an average value of 17 500 cal/mol is appropriate between 3·4° and 25 °C (Table 6). The computed relative rates at this value of $\bar{\mu}$ with a value of 33 000 cal/mol between 0 and 3·3 °C are

TABLE 5

APPARENT MEAN ACTIVATION ENERGIES ($\bar{\mu}$) FOR DIFFERENT FLESH TYPES
WITHIN EACH OF SIX ARBITRARILY CHOSEN TEMPERATURE RANGES

Temperature range and mean temperature	Flesh type	$\bar{\mu}$ (kcal/mol)	Sample range
$0°-3·3°$	Fish	32·6	75
	Poultry	36·3	3
	Beef	37·6	3
$3·4°-6·0°$	Fish	24·1	40
	Beef	21·4	9
$6·1°-10°$	Fish	21·8	20
	Beef	19·8	12
$10·1°-15°$	Fish	20·5	29
	Beef	17·5	4
$15·1°-20°$	Fish	17·1	27
	Beef	15·4	7
$20·1°-25°$	Fish	14·1	20

Data of Olley and Ratkowsky.[149]

close to the Olley and Ratkowsky[149] predictions up to 15 °C and fit the
observations of Daud et al.[41] at higher temperatures. These workers
showed a mean apparent activation energy of 18 600 cal/mol in 14
experimental systems and Morales[129] quoted 17 000 cal/mol as one of the
most common energy barriers in physiological processes and individual
enzyme systems.

TABLE 6

CALCULATION OF RELATIVE RATES TO GIVE ACTIVATION ENERGIES OF 15,
17·5, 20, 22·5 AND 25 kcal/mol FOR THE TEMPERATURE RANGE 3°–20 °C,
AND IN WHICH THE ACTIVATION ENERGY FOR THE RANGE 0°–3 °C IS
33 kcal/mol

Temperature (°C)	Relative rates for the following values of μ				
	15	17·5	20	22·5	25 kcal/mol
0	1	1	1	1	1
3	1·93	1·93	1·93	1·93	1·93
5	2·35	2·43	2·51	2·60	2·68
10	3·80	4·25	4·76	5·33	5·96
15	6·04	7·30	8·82	10·66	12·89
20	9·43	12·29	16·00	20·84	27·14

The majority of the data sets used by Olley and Ratkowsky[149] in the construction of their general spoilage curve were based on fish spoilage (Table 5). However, some data on meat and poultry were included and statistical analysis suggested that the relative spoilage rate was fairly constant for a given temperature, irrespective of the flesh type or the nature of the test (chemical, physical, bacteriological, organoleptic). Subsequently, Daud et al.[41] found that the rate of spoilage of chicken tissues and the rate of development and metabolism of poultry spoilage bacteria as a function of temperature were more accurately described by the Olley and Ratkowsky[149] predictions than by the Spencer and Baines[180] equation. However, it was noted that at temperatures above 16 °C the poultry spoilage curve continued to rise and did not show the plateau effect at a relative rate of c. 9 postulated by Olley and Ratkowsky (Fig. 6). Further data obtained in this laboratory suggests that this effect also occurs with fish and that the general spoilage curve may require modification in the range 15°–25 °C.[7] The continued rise in relative rate in the region 15°–25 °C may be expected from a consideration of the effect of temperature on the rate of growth of spoilage pseudomonads. Farrell and Barnes[50] noted generation times for a pseudomonad of 2·2 h at 15 °C, 1·4 h at 20 °C and 0·89 h at 25 °C. What remains to be proved is that an increased bacterial growth rate in this temperature range is reflected in a proportionately increased spoilage rate.

The Olley and Ratkowsky[149] curve was incorporated into the circuitry of an electronic temperature function integrator (Tefimupot, Solid State Equipment, PO Box 30-089, Lower Hutt, New Zealand) by Nixon.[140] This can be used to monitor the temperature history of the product and to provide an estimate of the equivalent number of days of 0 °C. From a knowledge of the expected shelf life of the product under a defined set of temperature conditions, the remaining shelf life can be calculated.

The use of the integrator in practice in the fishing industry has been described.[22,146,148,150] In this laboratory it has also been used to monitor temperature history during operations in a poultry processing plant thereby allowing the processor to estimate remaining shelf life during transport and retail storage.

It should be noted that, although the relative rates of spoilage of various substrates are similar, the actual time to reach a specified level will also depend on the nature of the substrates and the initial number of contaminants.[41] A continuing in-plant quality assurance programme is therefore necessary to maintain initial contamination levels at those expected by adherence to good manufacturing practice.

Temperature function integrators such as the Tefimupot therefore offer a means to monitor the temperature history and to predict the remaining shelf life of a food product. The general spoilage curve of Olley and Ratkowsky[149] refers to meats stored aerobically at high relative humidities (99–100 %). Deviation from these storage conditions will affect the relative spoilage rate, e.g. reduction in relative humidity[164] or an increase in CO_2[145] have greater effects at low temperatures than at high temperatures, decreasing k_0 and thus increasing the relative spoilage rate. Nevertheless, it is to be expected that electronic developments which will allow the production of smaller and less costly integrators will lead to considerably increased usage of these instruments to monitor the temperature history of aerobic, chill stored meats.

CONCLUSIONS

Jay[91] represented the problem of meat spoilage in a cyclical manner with interactions between the mechanisms of spoilage, detection of spoilage and a definition of spoilage. It was Jay's contention that 'a sound definition of meat spoilage seems vital to the proper elucidation of mechanisms and development of methods of detection'. In the interim period, no panacaeic definition of spoilage has emerged, nor, considering the inherent subjectivity of such a definition, is one likely to emerge. Indeed, it may be argued that Jay placed undue emphasis on this point and that methods for the detection of spoilage can only be derived from a detailed knowledge of the mechanisms of spoilage. While a precise knowledge of spoilage mechanisms has not yet been achieved, published work in the 1970s has provided valuable additional information and several promising methods for the evaluation of spoilage have emerged. Further studies of spoilage mechanisms should also consider the question of genetic control. Such studies have led to considerable progress in other areas of biodeterioration and could well provide novel means of meat preservation.

ACKNOWLEDGEMENTS

The author is indebted to Dr J. N. Olley, CSIRO, Division of Food Research, Hobart, and Dr C. J. Thomas of this Department, for helpful discussions during the preparation of this manuscript.

REFERENCES

1. ACKMAN, R. G., TOCHER, C. S., and McLACHLAN, J., *J. Fish. Res. Bd Can.*, 1966, **23**, 357.
2. ADAMCIC, M., and CLARK, D. S., *J. Food Sci.*, 1970, **35**, 103.
3. ADAMCIC, M., CLARK, D. S., and YAGUCHI, M., *J. Food Sci.*, 1970, **35**, 272.
4. ADAMS, R., FARBER, L., and LERKE, P., *Appl. Microbiol.*, 1964, **12**, 277.
5. ANDERSON, J. M., and BAIRD-PARKER, A. C., *J. Appl. Bacteriol.*, 1975, **39**, 111.
6. AYRES, J. C., *J. Appl. Bacteriol.*, 1960, **23**, 471.
7. BALL, A., B.Sc.Hons. Thesis, 1980, University of Tasmania, Hobart.
8. BARLOW, J., and KITCHELL, A. G., *J. Appl. Bacteriol.*, 1966, **29**, 188.
9. BARNES, E. M., *Roy. Soc. Health J.*, 1960, **80**, 145.
10. BARNES, E. M., *J. Sci. Food Agric.*, 1976, **27**, 777.
11. BARNES, E. M., and IMPEY, C. S., *J. Appl. Bacteriol.*, 1968, **31**, 97.
12. BARNES, E. M., IMPEY, C. S., GEESON, J. D., and BUHAGAIR, R. W. M., *Br. Poult. Sci.*, 1978, **19**, 77.
13. BARNES, E. M., and MELTON, W., *J. Appl. Bacteriol.*, 1971, **34**, 599.
14. BARNES, E. M., and SHRIMPTON, D. H., *J. Appl. Bacteriol.*, 1958, **21**, 313.
15. BARNES, E. M., and THORNLEY, M. J., *J. Food Technol.*, 1966, **1**, 113.
16. BEM, Z., HECHELMANN, H., and LEISTNER, L., *Fleischwirts.*, 1966, **56**, 985.
17. BLANKENSHIP, L. C., *Abstr. Ann. Meet. ASM*, 1977, **77**, 245.
18. BORTON, R. J., WEBB, N. B., and BRATZLER, L. J., *Food Technol.*, 1968, **22**, 94.
19. BORTON, R. J., BRATZLER, L. J., and PRICE, J. F., *J. Food Sci.*, 1970, **35**, 779.
20. BORTON, R. J., BRATZLER, L. J., and PRICE, J. F., *J. Food Sci.*, 1970, **35**, 783.
21. BROOKE, R. O., MENDELSOHN, J. M., and KING, F. J., *J. Fish Res. Bd Can.*, 1968, **25**, 2453.
22. BREMNER, H. A., OLLEY, J., and THROWER, S. J., *Tasmanian Regional Laboratory Occasional Paper No. 4.*, 1978, CSIRO.
23. BUTLER, J. L., STEWART, J. C., CARPENTER, Z. L., and VANDERZANT, C., *Abs. Ann. Meeting A.S.M.*, 1978, **78**, 186.
24. BUTLER, J. L., STEWART, J. C., VANDERZANT, C., CARPENTER, Z. L., and SMITH, G. C., *J. Food Prot.*, 1979, **42**, 401.
25. CASTELL, C. H., and GREENHOUGH, M. F., *J. Fish Res. Bd Can.*, 1957, **14**, 617.
26. CASTELL, C. H., and GREENHOUGH, M. F., *J. Fish Res. Bd Can.*, 1959, **16**, 21.
27. CERNY, G., *Europ. J. Appl. Microbiol.*, 1976, **3**, 223.
28. CERNY, G., *Europ. J. Appl. Microbiol. and Biotechnol.*, 1978, **5**, 113.
29. CHAI, T., CHEN, C., ROSEN, A., and LEVIN, R. E., *Appl. Microbiol.*, 1968, **16**, 1738.
30. CHALLENGER, F., and CHARLTON, P. T., *J. Chem. Soc.*, 1947, 424.
31. CHATAGNER, F., and SAURET-IGNAZI, G., *Bull. Soc. Chim. Biol.*, 1956, **38**, 415.
32. CLARK, D. S., *Poult. Sci.*, 1968, **49**, 1315.
33. CLARK, D. S., and LENTZ, C. P., *Can. Inst. Food Sci. Tech. J.*, 1973, **6**, 194.
34. CLARKE, P. H., *J. Gen. Microbiol.*, 1953, **8**, 397.
35. COWAN, S. T., and STEEL, K. J., *Identification of Medical Bacteria*, 2nd Ed. 1974, Cambridge University Press, London.
36. DAINTY, R. H., and HIBBARD, C. M., *J. Appl. Bacteriol.*, 1980, **48**, 387.

36   T. A. McMEEKIN

37. DAINTY, R. H., SHAW, B. G., DE BOER, K. A., and SCHEPS, R. G., *J. Appl. Bacteriol.*, 1975, **39**, 73.
38. DAINTY, R. H., SHAW, B. G., HARDING, C. D., and MICHANIE, S. *Cold Tolerant Microbes in Spoilage and the Environment*. eds. Russell, A. D., and Fuller, R., 1979, Academic Press, London.
39. DATEO, G., CLAPP, R., MACKAY, C., HEWITT, D. A. M., and HASSELSTRON, T., *Food Res.*, 1957, **22**, 440.
40. DAUD, H. B., McMEEKIN, T. A., and THOMAS, C. J., *Appl. Env. Microbiol.*, 1979, **37**, 399.
41. DAUD, H. B., McMEEKIN, T. A., and OLLEY, J., *Appl. Env. Microbiol.*, 1978, **36**, 650.
42. DUTSON, T. R., PEARSON, A. M., PRICE, J. F., SPINK, G. C., and TARRANT, P. J. V., *Appl. Microbiol.*, 1971, **22**, 1152.
43. EDDY, B. P., *J. Appl. Bacteriol.*, 1960, **23**, 189.
44. EDDY, B. P., and KITCHELL, A. G., *J. Appl. Bacteriol.*, 1959, **22**, 57.
45. ELLIOT, R. P., and MICHENER, H. D., *Technical Bulletin 1320*, 1965, USDA.
46. ELLMAN, G. L., *Arch. Biochem. Biophys.*, 1959, **82**, 70.
47. ELMOSSALAMI, E., and WASSEF, N., *Zbl. Vet. Med. B.*, 1971, **18**, 329.
48. EMPEY, W. A., and SCOTT, W. J., *C.S.I.R. Aust. Bull.*, 1939, 126.
49. ENFORS, S.-O., MOLIN, G., and TERNSTROM, A., *J. Appl. Bacteriol.*, 1979, **47**, 197.
50. FARRELL, A. J., and BARNES, E. M., *Br. Poult. Sci.*, 1964, **5**, 89.
51. FIRSTENBERG-EDEN, R., NOTERMANS, S., and VAN SCHOTHORST, M. *J. Food Safety*, 1978, **1**, 217.
52. FIRSTENBERG-EDEN, R., NOTERMANS, S., THIEL, F., HENSTRA, S., and KAMPELMACHER, E. H., *J. Food Prot.*, 1979, **42**, 305.
53. FREEMAN, L. R., SILVERMAN, G. J., ANGELINI, P., MERRIT, C., and ESSELEN, W. B., *Appl. Env. Microbiol.*, 1976, **32**, 222.
54. FROMM, D., and MONROE, R. J., *Poult. Sci.*, 1965, **44**, 325.
55. GARDNER, G. A., Ph.D. Thesis, 1965, Queen's University, Belfast.
56. GARDNER, G. A., CARSON, A. W., and PATTON, J., *J. Appl. Bacteriol.*, 1967, **30**, 321.
57. GARDNER, G. A., and PATTERSON, R. L. S., *J. Appl. Bacteriol.*, 1975, **39**, 263.
58. GARDNER, G. A., and STEWART, D. J., *J. Sci. Fd Agric.*, 1966, **17**, 491.
59. GIBBS, P. A., PATTERSON, J. T., and HARPER, D. B., *J. Sci. Fd Agric.*, 1979, **30**, 1109.
60. GILL, C. O., *J. Appl. Bacteriol.*, 1976, **41**, 401.
61. GILL, C. O., and NEWTON, K. G., *J. Appl. Bacteriol.*, 1977, **43**, 189.
62. GILL, C. O., and NEWTON, K. G., *Meat Science*, 1978, **2**, 217.
63. GILL, C. O., and NEWTON, K. G., *Appl. Env. Microbiol.*, 1979, **37**, 362.
64. GILL, G. O., and PENNEY, N., *Appl. Env. Microbiol.*, 1977, **33**, 1284.
65. GILLESPIE, N. C., *Reported Proceedings of the Australian Fish Exposition*, Melbourne, 1976, Australian Government Publishing Service, Canberra.
66. GILLILAND, S. E., and SPECK, M. L., *J. Food Sci.*, 1975, **40**, 903.
67. GRAY, G. S., and MILLER, C. A., *Hlth. Lab. Sci.*, 1978, **15**, 150.
68. GUCKERT, A., BREISCH, A., and REISINGER, O., *Soil Biol. Biochem.*, 1975, **7**, 241.
69. HAIGHT, R. D., and MORITA, R. Y., *J. Bacteriol.*, **92**, 1388.

70. HALLECK, F. E., BALL, C. O., and STIER, E. F., *Food Technol.*, *Champaign*, 1958, **12**, 197.
71. HANUS, F. J., and MORITA, R. Y., *J. Bacteriol.*, 1968, **95**, 736.
72. HARDER, W., and VELDKAMP, H., *Antonie van Leeuwenhoek*, 1971, **37**, 51.
73. HASEGAWA, T., PEARSON, A. M., PRICE, J. F., and LECHOVICH, R. V., *Appl. Microbiol.*, 1970, **20**, 117.
74. HASEGAWA, T., PEARSON, A. M., PRICE, J. F., RAMPTON, J. H., and LECHOVICH, R. V., *J. Food Sci.*, 1970, **35**, 720.
75. HECHELMANN, H., BEM, Z., UCHIDA, K., and LEISTNER, L., *Fleischwirts*, 1974, **54**, 1515.
76. HERBERT, R. A., Ph.D. Thesis, 1970, University of Aberdeen.
77. HERBERT, R. A., HENDRIE, M. S., GIBSON, D. M., and SHEWAN, J. M., *J. Appl. Bacteriol.*, 1971, **34**, 41.
78. HERBERT, R. A., ELLIS, R. J., and SHEWAN, J. M., *J. Sci. Fd Agric.*, 1975, **26**, 1187.
79. HERBERT, R. A., and SHEWAN, J. M., *J. Sci. Fd Agric.*, 1975, **26**, 1195.
80. HERBERT, R. A., and SHEWAN, J. M., *J. Sci. Fd Agric.*, 1976, **27**, 89.
81. HUNTER, A. C., and LINDEN, B. A., *Am. Food J.*, 1923, **18**, 538.
82. INGRAHAM, J. L., *J. Bacteriol.*, 1958, **76**, 75.
83. INGRAM, M., *Proc. Soc. Appl. Bacteriol.*, 1951, **14**, 243.
84. INGRAM, M., and DAINTY, R. H., *J. Appl. Bacteriol.*, 1971, **34**, 21.
85. INGRAM, M., and MACKEY, B. M., *Inhibition and Inactivation of Vegetative Microbes*, eds. Skinner, F. A., and Hugo, W. B., 1976, Academic Press, London.
86. JAY, J. M., *Food Technol.*, 1964, **18**, 1633.
87. JAY, J. M., *Food Technol.*, 1964, **18**, 1637.
88. JAY, J. M., *The Physiology and Biochemistry of Muscle as a Food*, eds. Briskey, E. J., Cassens, R. G., and Trautman, J. C., 1966, Univ. Wisconsin Press, Madison.
89. JAY, J. M., *Hlth. Lab. Sci.*, 1966, **3**, 101.
90. JAY, J. M., *Appl. Microbiol.*, 1966, **14**, 492.
91. JAY, J. M., *J. Milk Food Technol.*, 1972, **35**, 467.
92. JAY, J. M., *J. Appl. Bacteriol.*, 1977, **43**, 99.
93. JAY, J. M., *Modern Food Microbiology*, 2nd Ed., 1978, D. Van Nostrand Co. New York.
94. JAY, J. M., and KONTOU, K. S., *Appl. Microbiol.*, 1967, **15**, 759.
95. JAYE, J. M., KITTAKE, R. S., and ORDAL, Z. J., *Food Technol.*, 1962, **16**, 95.
96. JOCELYN, P. C., *Biochemistry of the -SH group*, 1972, Academic Press, London.
97. KADOTA, H., and ISHIDA, Y., *Ann. Rev. Microbiol.*, 1972, **26**, 127.
98. KALLIO, R. H., and LARSON, A. D., Amino Acid Metabolism, eds. McElroy, W. D., and Glass, B., 1955, Johns Hopkins Press, Baltimore, Maryland.
99. KNOOP, G. B., PARMALEE, C. E., and STADELMAN, W. J., *Poult. Sci.*, 1971, **50**, 530.
100. KRAFT, A. A., OBLINGER, J. L., WALKER, H. W., KAWAL, M. C., MOON, N. J., and REINBOLD, G. W., *Microbiology in Agriculture, Fisheries and Food*, eds. Skinner F. A., and Carr, J. G., 1976, Academic Press, London.
101. LAAKSO, S., *J. Gen. Microbiol.*, 1976, **95**, 391.

102. LAHELLEC, C., MEURIER, C., and CATSARAS, M., *Ann. Rech. Vet.*, 1972, 3, 421.
103. LAPIN, R. M., and KOBURGER, J. A., *Appl. Microbiol.*, 1974, 27, 666.
104. LATHAM, M. J., BROOKER, B. E., PETTIPHER, G. L., and HARRIS, P. J., *Appl. Env. Microbiol.*, 1978, 35, 156.
105. LAW, B. M., and SHARPE, M. E., *J. Dairy Res.*, 1978, 45, 267.
106. LAWRIE, R. A., *Meat Science* 3rd Ed., 1979, Pergamon Press, London.
107. LEA, C. H., STEVENS, B. J. H., and SMITH, M. J., *Br. Poult. Sci.*, 1969, 10, 203.
108. LEE, J. V., and SHEWAN, J. M., *J. Gen. Microbiol.*, 1977, 98, 439.
109. LEISTNER, L. and RODEL, W., *Inhibition and Inactivation of Vegetative Microbes*, Eds. Skinner, F. A., and Hugo, W. B., 1976, Academic Press, London.
110. LERKE, P., FARBER, L., and ADAMS, R., *Appl. Microbiol.*, 1967, 15, 777.
111. LEVIN, R. E., *Appl. Microbiol.*, 1968, 16, 1734.
112. LONG, H. F., and HAMMER, B. W., *Res. Bull. Iowa Agric. Exp. Stn*, 1941, 285, 176.
113. McCOWAN, R. P., CHENG, K. J., BAILEY, C. B. M., and COSTERTON, J. W., *Appl. Env. Microbiol.*, 1978, 35, 149.
114. McMEEKIN, T. A., *Appl. Microbiol.*, 1975, 29, 44.
115. McMEEKIN, T. A., *Appl. Env. Microbiol.*, 1977, 33, 1244.
116. McMEEKIN, T. A., GIBBS, P. A., and PATTERSON, J. T., *Appl. Env. Microbiol.*, 1978, 35, 1216.
117. McMEEKIN, T. A., and PATTERSON, J. T., *Appl. Microbiol.*, 1975, 29, 163.
118. McMEEKIN, T. A., and THOMAS, C. J., *J. Appl. Bact.*, 1978, 45, 383.
119. McMEEKIN, T. A., THOMAS, C. J., and McCALL, D., *J. Appl. Bact.*, 1979, 46, 195.
120. MANNING, D. J., *J. Dairy Res.*, 1974, 41, 81.
121. MANNING, D. J., CHAPMAN, H. R., and HOSKING, L. D., *J. Dairy Res.*, 1976, 43, 313.
122. MANNING, D. J., *J. Dairy Res.*, 1979, 46, 523.
123. MEERS, J. L., *CRC Critical Reviews in Microbiology*, 1974, 139.
124. MILLER, III, A., SCANLAN, R. A., LEE, J. S., LIBBEY, L. M., and MORGAN, M. E., *Appl. Microbiol.*, 1973, 25, 257.
125. MILLER, III, A., SCANLAN, R. A., Lee, J. S., and LIBBEY, L. M., *Appl. Microbiol.*, 1973, 25, 952.
126. MILLER, III. A., SCANLAN, R. A., LEE, J. S., and LIBBEY, L. M., *Appl. Microbiol.*, 1973, 26, 18.
127. MINOR, L. J., PEARSON, A. M., DAWSON, L. E., and SCHWEIGERT, B. S., *J. Food Sci.*, 1965, 30, 686.
128. MINOR, L. J., PEARSON, A. M., DAWSON, L. E., and SCHWEIGERT, B. S., *J. Agric. Food Chem.*, 1965, 13, 298.
129. MORALES, M. F., *J. Cell Comp. Physiol.*, 1947, 30, 303.
130. MORITA, R. Y., *Bact. Rev.*, 1975, 39, 144.
131. MOSSEL, D. A. A., *J. Appl. Bacteriol.*, 1971, 34, 96.
132. MOTOHIRO, T., *Mem. Fac. Fish. Hokkaido Univ.*, 1962, 10, 1.
133. NEAL, J. J. and BANWART, G. J., *J. Food Sci.*, 1977, 42, 555.
134. NEWTON, K. G., and GILL, C. O., *Appl. Env. Microbiol.*, 1978, 36, 375.
135. NEWTON, K. G., and GILL, C. O., *J. Appl. Bacteriol.*, 1978, 44, 91.
136. NEWTON, K. G., and GILL, C. O., *Appl. Env. Microbiol.*, 1979, 36, 375.

137. NEWTON, K. G., HARRISON, J. C. L., and SMITH, K. M., *J. Appl. Bacteriol.*, 1977, **45**, 53.
138. NEWTON, K. G., HARRISON, J. C. L., and WAUTERS, A. M., *J. Appl. Bacteriol.*, 1978, **45**, 75.
139. NICOL, D. J., SHAW, M. K., and LEDWARD, D. A., *Appl. Microbiol.*, 1970, **19**, 93.
140. NIXON, P. A., '*Report on Quality in Fish Products*', 1971, Fishing Industry Board, Wellington, New Zealand.
141. NOTERMANS, S., FIRSTENBERG-EDEN, R., and VAN SCHOTHORST, M., *J. Food Prot.*, 1979, **42**, 228.
142. NOTERMANS, S., and KAMPELMACHER, E. H., *Br. Poult. Sci.*, 1974, **15**, 573.
143. NOTERMANS, S., and KAMPELMACHER, E. H., *Br. Poult. Sci.*, 1975, **16**, 487.
144. NOTERMANS, S., VAN LEUSDEN, F. M., and VAN SCHOTHORST, M., *J. Appl. Bacteriol.*, 1977, **43**, 383.
145. OGILVY, W. S., and AYRES, J. C., *Food Technol.*, 1951, **5**, 97.
146. OLLEY, J., *Bull. Int. Inst. Froid. Commisions C2, D1, D2, D3, E1 Melbourne, Australia*, 1976, International Institute of Refrigeration, Paris.
147. OLLEY, J., *Int. J. Refrig.*, 1978, **1**, 81.
148. OLLEY, J., DAUD, H. B., and McMEEKIN, T. A., *IPFC Symposium on Fish Utilisation Technology and Marketing in the IPFC Region*. 1978, Manila, Philippines.
149. OLLEY, J., and RATKOWSKY, D. A., *Food Technol. Aust.*, 1973, **25**, 66.
150. OLLEY, J., and RATKOWSKY, D. A., *Food Technol. NZ*, 1973, **8**, 13.
151. PARR, L. J., and LEVETT, G., *J. Food Technol.*, 1969, **4**, 283.
152. PATTERSON, J. T., and GIBBS, P. A., *J. Appl. Bacteriol.*, 1977, **43**, 25.
153. PATTERSON, J. T., and GIBBS, P. A., *J. Food Technol.*, 1978, **13**, 1.
154. PATTERSON, J. T., and GIBBS, P. A., *Meat Sci.*, 1978, **2**, 263.
155. POTTINGER, S. R., *Comm. Fisheries Rev.*, 1948, **10**, 1.
156. RACCACH, M., and BAKER, R. C., *J. Food Prot.*, 1978, **41**, 703.
157. RACCACH, M., BAKER, R. C., REGENSTEIN, J. M., and MULNIX, E. J., *J. Food Sci.*, 1979, **44**, 43.
158. RICHARDS, J. C. S., JASON, A. C., HOBBS, G., GIBSON, D. M., and CHRISTIE, R. H., *J. Phys. E. Sci. Instrum.*, 1978, **11**, 560.
159. RONALD, A. P., and THOMSON, W. A. B., *J. Fish Res. Bd Can.*, 1964, **21**, 1481.
160. RONSIVALLI, L. J., and CHARM, S. E., *Mar. Fish. Rev.*, 1975, **37**, 32.
161. ROTH, L. A., and CLARK, D. S., *Can. J. Microbiol.*, 1975, **21**, 629.
162. SAFFLE, R. L., MAY, K. N., HAMID, H. A., and IRBY, J. D., *Fd Technol., Champaign*, 1961, **15**, 465.
163. SANDINE, W. E., MURALIDHARA, K. S., ELLIKER, P. R., and ENGLANG, D. C., *J. Milk Food Technol.*, 1972, **35**, 691.
164. SCOTT, W. J., *J. Counc. Sci. Ind. Res. (Aust.)*, 1937, **10**, 338.
165. SCOTT, W. J., *J. Counc. Sci. Ind. Res. (Aust.)*, 1938, **11**, 266.
166. SEGAL, W., and STARKEY, R. L., *J. Bacteriol.*, 1969, **98**, 908.
167. SHARPE, M. E., LAW, B. A., and PHILLIPS, B. A., *J. Gen. Microbiol.*, 1976, **94**, 430.
168. SHARPE, M. E., LAW, B. A., PHILLIPS, B. A., and PITCHER, D. G., *J. Gen. Microbiol.*, 1977, **101**, 345.
169. SHAW, M. K., *J. Bacteriol.*, 1967, **93**, 1332.

170. SHAW, M. K., and NICOL, D. J., *Proc. 15th Europ. Meeting of Meat Res. Workers*, 1969, 226.
171. SHELEF, L. A., *J. Appl. Bacteriol.*, 1975, **39**, 273.
172. SHELEF, L. A., *J. Food Sci.*, 1977, **42**, 1172.
173. SHELEF, L. A., and JAY, J. M., *Appl. Microbiol.*, 1969, **17**, 931.
174. SHEWAN, J. M., *J. Appl. Bacteriol.*, 1971, **34**, 299.
175. SHEWAN, J. M., and MURRAY, C. K., *Cold Tolerant Microorganisms in Spoilage and the Environment*, eds. Russell, A. D., and Fuller, R., 1979, Academic Press, London.
176. SHRIMPTON, D. H. and BARNES, E. M., *Chemy. Ind.*, 1960, 1492.
177. SIEBURTH, J. McN., *J. Bacteriol.*, 1959, **77**, 521.
178. SIKES, A., and MAXCY, R. B., *J. Food Sci.*, 1979, **44**, 1228.
179. SIPOS, J. C., and ACKMAN, R. G., *J. Fish Res. Bd Can.*, 1964, **21**, 423.
180. SPENCER, R., and BAINES, C. R., *Food Technol.*, 1964, **18**, 175.
181. TARRANT, P. J. V., PEARSON, A. M., PRICE, J. F., and LECHOVICH, R. V., *Appl. Microbiol.*, 1971, **22**, 224.
182. TARRANT, P. J. V., JENKINS, N., PEARSON, A. M., and DUTSON, T. R., *Appl. Microbiol.*, 1973, **25**, 996.
183. TAYLOR, A. A., and SHAW, B. G., *J. Food Technol.*, 1977, **12**, 515.
184. THOMAS, C. J., Ph.D. Thesis, 1979, University of Tasmania, Hobart.
185. THOMAS, M., *Mon. Bull. Minist. Hlth Public Hlth Lab. Serv.*, 1966, **25**, 42.
186. TROLLER, J. A., and CHRISTIAN, J. H. B., *Water Activity and Food*, 1978, Academic Press, New York.
187. WELLS, F. E., SPENCER, J. V., and STADELMANN, W. J., *Food Technol.*, 1958, **12**, 425.
188. WIEDMANN, J. F., *J. Appl. Bacteriol.*, 1965, **28**, 365.

Chapter 2

THE NURMI CONCEPT AND ITS ROLE IN THE CONTROL OF SALMONELLAE IN POULTRY

H. Pivnick

Department of National Health and Welfare, Ottawa, Canada

and

E. Nurmi

State Veterinary Medical Institute, Helsinki, Finland

SUMMARY

Salmonella *infection of poultry is of little economic consequence to the poultry industry but is the source of widespread food-borne salmonellosis in man. The causes of infection in poultry are infected breeders, contaminated feed and, to a lesser extent, a contaminated environment. Elementary and simple measures would be adequate for control, but are applied uniformly in only a few countries.*

The introduction of gut content from adult birds, or cultures of gut content, into newly hatched chicks and turkey poults makes them resistant to over 10^3 infectious doses of Salmonella. *However, such treatment is not a panacea; conventional methods of control of breeder birds and feed are also necessary to bring about and maintain a decrease in the incidence of* Salmonella-*contaminated poultry meat.*

The succession of microorganisms in the gut of the chicken is an orderly process that takes place over several weeks. The immediate protective effect against Salmonella (*within one hour of introducing caecal content from an adult bird into a newly hatched chick*) *suggests that the protective mechanism may be physical. One hypothesis is that the glycocalyces of the protective bacteria bind to the glycocalyces of specific gut epithelial cells thus excluding* Salmonella *from the sites they need for invasion. To date, there appear to be no pure cultures of bacteria that are equal to the mixed caecal or faecal flora of mature poultry for preventing infection of newly hatched poultry by* Salmonella.

INTRODUCTION

The Nurmi concept for diminishing infection of poultry by salmonellae involves *per ora* introduction of gastrointestinal flora from adult birds into newly hatched chicks or poults. This treatment makes them immediately resistant to 10^3–10^6 infectious doses of salmonellae.[62,76] One of us (H.P.) has given the name, Nurmi concept, to this discovery.[35]

The Nurmi concept may apply to flocks for producing fertile eggs, a mass production system for hatching the eggs, and a growing system for rearing newly hatched chicks to the stage suitable for meat or laying eggs. The concept appears to be applicable to the control of *Salmonella* in turkeys, but there has been no investigation of its applicability to ducks, geese, domesticated pheasants or other poultry which are minor sources of meat. The concept is especially applicable to commercial flocks.

The concept enables the introduction of a novel method for decreasing infection of poultry by *Salmonella*. The method will probably not eliminate infection by *Salmonella* however, unless complementary measures for control are also taken, e.g. use of disease-free breeding stock, *Salmonella*-free feed and effective control of *Salmonella* in the environment.[58]

Salmonella spp. can infect all animals produced for meat; infection is almost always *per os*, usually via feed or water. The salmonellae can be divided into two main groups: the host-adapted and the non-host-adapted. The host-adapted, which constitute less than 1 % of approximately 1800 serovars, usually infect only one or a few animal species; in contrast, the non-host-adapted may infect many animal species. Thus, the host-adapted *Salmonella typhi* infects humans often with fatal consequences, but does not infect poultry. Similarly, the host-adapted *Salmonella pullorum* and *Salmonella gallinarum* infect poultry, often with fatal consequences, but infect humans only rarely. The non-host-adapted serotypes, however, only occasionally cause fatal disease of either man or other animals and are, therefore, subjected to less stringent measures for control. Of the 30 non-host-adapted serovars commonly found in poultry meat, only 10 are responsible for most cases of salmonellosis in man.[70] The control of

Salmonella in poultry flocks is, therefore, a major concern of public health agencies but a minor concern of agriculturists.

ECONOMIC CONSIDERATIONS

Poultry
Host-adapted
S. pullorum and *S. gallinarum* are so lethal to young chicks that their control was essential for the development of the modern poultry industry. They caused fatal illness of young chicks, the fatality rate often exceeding 25 % of the flock. As poultry raising evolved from the small family-owned operation to the present system of mass production, losses of 25 % were economically intolerable. Thus, breeding flocks that supplied fertile eggs to factory-type hatcheries were controlled for *S. pullorum* and *S. gallinarum*. In countries with modern poultry production, these serovars are rarely found.[108]

Non-host-adapted
The non-host-adapted serovars rarely cause economically disruptive losses to the poultry industry: they account for less than 1 % of diagnosed poultry diseases.[108] They may, however, occasionally cause losses up to 25 % in individual operations.[43] These high losses may be related to concomitant stresses such as intercurrent infections with other sublethal agents and adverse high or low temperatures.

Man
Costs
Epidemiological investigation and reporting of human salmonellosis varies considerably between countries. Thus, the following data must be considered in that context. The reported cases per 100 000 population in 1978 have been calculated from various sources, and are approximate: Canada, 37; USA, 14; England, Wales and Northern Ireland, 20; Denmark, 10; Finland, 44 and Sweden, 43.[61] In the last three countries from one-third to two-thirds of infections occur during foreign travel. There are an estimated 30 cases of human salmonellosis for each case that is reported through official national agencies in Canada and the United States.[38] Thus, the true incidence, at least in Canada and the USA, is about 30 times higher than reported, and about 1 % and 0·4 % of the populations, respectively, are infected annually.

The cost per case in Canada and the USA is about $500, the loss of productivity exceeding direct medical costs. Estimated costs are about $100 000 000 per year for Canada, and over $1 000 000 000 for the USA.[110] Thus, the cost of human salmonellosis is considerable.

Deaths from human salmonellosis caused by non-host-adapted serovars in countries with advanced medical treatment are rare, usually $<0.3\%$ of reported cases. Thus, a disease which causes minor economic losses to the poultry industry and only a few deaths in the human population usually cannot be given the highest priority by either agricultural or public health interests.

Poultry Meat as a Source of Infection For Man
There is a paucity of reliable data, based on statistically designed sampling procedures and uniform methodology, that gives the incidence of *Salmonella*-contaminated poultry carcasses. A recent review[110] indicates that the incidence is usually in the range of 25–65%, but lower in some countries with less than 1% in Sweden,[4] about 5% in Denmark and about 7% in Finland.[72,110] However, we emphasise that different sampling plans and methodologies exist between countries. The question is often asked whether this decreased incidence of *Salmonella*-contaminated poultry in Scandinavian countries has resulted in a decreased incidence of human salmonellosis in these countries. There is no useful answer: the per capita consumption of poultry in Scandinavian countries is only 20–30% of that in Canada and the USA.[110] Also, there are only limited data that directly relate human salmonellosis with the consumption of poultry meat. There are, however, many well documented reports of the distribution of *Salmonella* from contaminated feed to poultry with subsequent increases of salmonellosis in the human population.[22,64,112] In these outbreaks, introduction in poultry feed of a serotype previously not common in a country and subsequent discovery of the same serotype in poultry flocks, poultry meat and the human population of that country has provided epidemiological evidence that appears irrefutable. Moreover, there is a high correlation between the serotypes most frequently found in poultry and those causing infection of humans.[70]

CAUSES OF SALMONELLA INFECTION OF POULTRY

Environment
Salmonella in barns and machinery, and from infected insects, wild birds and rodents, may infect chicks. The extent of this source of infection is, however, probably minor.

Breeders

Breeding hens that are *Salmonella*-infected lay fertile eggs that are rarely infected prior to deposition of the shell, but are frequently contaminated with *Salmonella* on the outside of the shell. *Salmonella* can penetrate the shell given adequate moisture and a differential in temperature between the exterior and interior of the egg. The infected fertile egg may die prior to hatching, may hatch into a weak chick that is culled, or worse, develop into a healthy chick[49] that excretes up to 10^8 cells of *Salmonella* per gram of faeces;[62] spread of *Salmonella* from an infected chick to its cohorts may occur in the hatchery[25,36] and be rapid and widespread when infected chicks are reared on litter.[99] Penetration of the shell by *Salmonella* can be prevented if the eggs are gathered frequently and fumigated promptly with formaldehyde.[36] This preventive measure is, however, applied uniformly in only a few countries.

Feed

Feed for poultry is usually compounded from protein rich material and cereal with other nutrients in smaller amounts. The protein component is usually rendered from inedible material from slaughtered animals, animals unfit for human consumption and fish, or it is the residue from seeds extracted to obtain edible oil. Although these materials are processed under conditions which destroy *Salmonella*, sanitation and other elementary preventive measures are sometimes inadequate and the finished products especially those derived from livestock and poultry, are often recontaminated with *Salmonella*. Thus, although the breeder and hatchery operations may produce *Salmonella*-free chicks and poults, they may be infected by *Salmonella*-contaminated feed.[116]

Programmes to reduce *Salmonella*-contamination of feed have been effective in only a few countries, e.g. Denmark, Sweden and more recently, Finland. In most other countries, there appears to be no economic benefit to renderers who produce a *Salmonella*-free product. Nevertheless, a substantial number of renderers appear to be able to consistently produce *Salmonella*-free protein concentrates.[115]

The low incidence of *Salmonella*-contaminated poultry meat in Sweden and Denmark is largely due to the regulation of feed and quarantine of imported breeder stock. However, in countries that do not regulate feed for *Salmonella*, there is little incentive for poultry producers to purchase *Salmonella*-free feed. Nor have departments of agriculture in most countries placed a high priority on decreasing *Salmonella* in poultry, when diseases of large animals and some other poultry diseases are of greater, and sometimes enormous, importance to national economies.[23] Thus, there is clearly a need for some other method to decrease *Salmonella* in poultry. Application of the Nurmi concept may fulfil some of this need even if it does not solve all the difficulties.

EXCLUSION OF SALMONELLA BY NORMAL GUT MICROBIOTA

Infection

Infection by *Salmonella*, with rare exceptions, is *per os*. *Salmonella* appear able to enter the mucosa in any section of the gut, but they persist longer in the caeca, caecal tonsils and large intestine.[28] In mice, establishment in the caeca appears to be a requirement for sustained infection,[51] although Peyers patches in the distal ileum are penetrated before those in the caecum.[20] Penetration of caecal epithelium begins with contact between glycocalyces on the *Salmonella* and glycocalyces on the microvilli of the brush border, destruction of the brush border and terminal web, and evagination of mucosal cells;[109,111] penetration through intercellular junctions occurs less frequently.[109]

Exclusion

The ability of the autochthonous (indigenous) gut microflora to protect mice against *per ora* infection by *Salmonella* has been known for several decades. Disruption of the gut flora with antibiotics decreased the ID_{50} to about $1/10\,000$. Repopulation of the gut with faeces of normal mice restored resistance gradually, increasing the *per os* ID_{50} about 10 fold within 6 h and about 100 fold within 18 h.[14,16,53,54]

There are several mechanisms by which the autochthonous flora prevent infection.[90] Two of these appear pertinent to our discussion: (a) production of volatile fatty acids (VFA) in the caeca and (b) occupation of sites on the mucosa that *Salmonella* invade.

The main VFA produced by obligate anaerobes in the caeca are, in order of concentration, acetic acid, butyric acid and propionic acid in mice[15,16,51,52] and chickens.[10] In the chicken, these three acids are produced maximally by the second week after hatching. The acids, in concentrations produced *in vivo* in the caeca of both mice and chickens inhibit growth of *Salmonella in vitro* if the pH is adjusted to that of the caeca (about 6·0).[10,16] The E_h ($-0\cdot2$ to $-0\cdot4$ V) in the mouse caeca contributes to the inhibitory effect of the VFA.[51] The effect is bacteriostatic, and probably not bacteriocidal.[52] The kinetics of passage through the gut favour elimination of bacteria experiencing bacteriostasis.[52]

Some of the autochthonous flora may adhere to sites that *Salmonella* require for invasion.[102] The rapidity of the protective effect, within one hour of introducing adult faecal flora to chicks,[93] suggests that adhering bacteria may exclude *Salmonella*, since sufficient VFA would not be produced so rapidly.

Adhering bacteria are defined as those that are not readily removed from the epithelium by specified washing procedures. Many have been cultured, but some observed *in situ* apparently cannot be cultured by methods presently available. Numerous species of adhering microorganisms have been found on the epithelia of a wide variety of intestinal tissues in many species of animals.[88,92] They appear to have specific functions that benefit the animal.[21,24] Two of these, scavenging of oxygen to permit establishment of VFA-producing obligate anaerobes in the caeca[51] and the above mentioned competitive exclusion of *Salmonella*,[47,48] appear to be important in the context of this review.

There appear to be several methods whereby different bacteria adhere[88,89,91] and all may be biologically important to the host. Most attention, however, has been focused on the glycocalyces. Both bacteria and animal cells produce a glycocalyx (a mass of long, charged poly-saccharide fibres) and the glycocalyx of the animal cell bonds with that of the bacteria, possibly through divalent cations.[24] Bacteria which mutate with a loss of ability to synthesise a glycocalyx also lose their ability to adhere. The significance of this loss will be discussed later.

In the chicken gut, adhering bacteria are plentiful in the crop, ileum and caecum and are less plentiful or absent from other areas.[34] Lactobacilli dominate in the crop[18,30,32] and obligately anaerobic Gram-positive cocci in the caecum.[34] The possible protective activity of adhering lactobacilli against enteric pathogens was recognised by Fuller[30] in 1973 who suggested dosing newly hatched chicks with chicken faeces or pure cultures of adhering lactobacilli. In the same year, Nurmi and Rantala[62] dosed chicks with caecal material from adult chickens, thus preventing infection by *Salmonella*.

THE NURMI CONCEPT INVOLVES THE FOLLOWING POINTS

(a)   Newly hatched chicks may be infected by a single cell of *Salmonella*.
(b)   Older birds are resistant to infection because of the autochthonous

microbiota of the gut, particularly the caeca and colon, but possibly other portions of the gut.

(c) Chicks hatched by sitting hens are probably populated rapidly by the autochthonous gut microflora of the adult.

(d) Hatcheries have replaced sitting hens, however, and the mass production of chicks is carried out in such a sanitary environment that the autochthonous microflora is not introduced at the modern hatchery.

(e) The rearing barns in which the newly hatched chicks are placed are usually sanitised and the floors covered with fresh litter between crops of chickens. Thus, autochthonous flora of adult birds are not readily available to populate the gut of the newly hatched chicks.

(f) The introduction of intestinal microflora of adult birds to newly hatched chicks makes most of them immediately resistant to 10^3–10^6 infectious doses of *Salmonella*.

(g) The intestinal flora of adult birds may be introduced as a suspension of faecal droppings, caecal material or anaerobic cultures; these are designated as 'treatments'. Treatments may be introduced directly into the crop, or by addition to the drinking water and possibly to the feed. Aerosols may also be useful vectors.

(h) The source of the treatment is the homologous species, although treatment derived from chicken protects turkeys and vice versa.

RECIPROCAL PROTECTION

A single source of adult chickens can provide gastrointestinal material that is protective for a wide variety of breeds and strains of chicks of both sexes.[100] Adult turkeys protect poults,[13,48,105] but there is no information on protection between breeds or strains of turkeys. Chickens can protect turkey poults[100] and turkeys can protect chicks.[105] Ducks protect chicks,[105] but faecal material from seven other species of birds did not protect chicks.[101] Gastrointestinal contents of other animals (sheep, rat, horse, cow) did not protect chicks.[77,105]

DEFINITION OF TERMS

Donor
A bird, usually mature, used as a source of gastrointestinal material.

Suspension

A suspension of faecal, caecal or other gastrointestinal material that may be administered to chicks or poults, or may be inoculated into sterile medium to initiate a culture.

Culture

Growth in a culture medium inoculated with a suspension.

Treatment

Introduction of suspension or culture into the desired bird by direct addition to the crop or in the drinking water. Treatment should be given as soon as possible after hatching, preferably at the hatchery,[103] to minimise spread of *Salmonella* from an occasional chick infected prior to hatching. Older *Salmonella*-infected birds may be treated with a suspension or culture subsequent to antibiotic therapy; this will be discussed later.

Efficacy

The ability of treated birds to withstand infection compared with untreated control birds. The idealised concept of efficacy is complete protection against unlimited numbers of all serotypes with the possible exceptions of *S. pullorum* and *S. gallinarum*. The ideal will not be realised, however, and a more realistic approach is to define an effective treatment as that which decreases the number of infected chicks to a given percentage (e.g. $< 10\%$ or $< 25\%$) of the incidence of infection in non-treated control birds challenged and housed under identical conditions. The time after treatment at which the comparison is made could vary depending on the destiny of the bird, but for practical purposes, a period of 7–14 days appears adequate as normally the excretion of *Salmonella* following infection is highest during this period.[19] The objective of treatment is not to prevent all infection, but to minimise spread to pen mates from those chicks that do become infected. Chickens, even without treatment, become resistant to infection by moderate doses of *Salmonella* within 2–4 weeks of hatching. There is a need to introduce quantitation into the definition of efficacy to facilitate comparison of treatments.

SOURCES OF MATERIAL FOR TREATMENT

Live Birds

There are some reports that mature chickens from some sources are donors of more protective cultures than mature chickens from other sources.

However, there is little or no accompanying data that is statistically significant to support these claims[17,102,105] and this area needs considerable further investigation.

The age at which a bird, held under normal conditions, becomes a donor of fully protective bacteria is unknown. The microbial populations in the chicken gut differ with age, but under normal conditions of rearing appear to climax at about 3–5 weeks of age.[11,65] Presumably treatment speeds up, or even eliminates the need for autogenic and allogenic forces that direct the successions of populations required for a day-old chick to achieve the autochthonous flora of the adult gut; or perhaps the protective bacteria are only a few of the many species that form the climax population.[3] One report[17] indicates that day-old chicks given faecal flora of adult birds became, within three days of treatment, donors of fully protective flora.Their gut microflora, given to day-old chicks gave the same protection against infection as gut flora of adult birds. However, most investigators of the Nurmi concept have used adult chickens as donors.

Flocks of specific pathogen free (SPF) chickens may be produced to serve as donors.[101] Day-old chicks treated with twice-passaged cultures derived from faeces of mature birds, were held in isolation. Testing over a year found that the flock was free of specified poultry pathogens, including some viruses. Faeces from this flock were fully protective.

Portion of Intestinal Tract

Different segments of the donor's intestinal tract are not equally suitable as sources of material for treatment. Lumen content from caeca or the large intestine appears to be the most protective.[48,62] Faeces, which may contain bacteria from all portions of the tract, give good protection. Some macerated mucosal tissues may contain a high percentage of adhering bacteria but mucosa from the crop is not effective.[100] To date, however, there have been insufficient data published to establish with certainty the suitability of bacteria in other portions of the tract for protecting chicks or poults.

Preservation of Material for Treatment and for Inoculation

Preserved material has been used in two ways: (a) for direct treatment of chicks and (b) to inoculate cultures. Material may be frozen, refrigerated and lyophilised, but there is little detailed information on techniques for the preferred method of preservation or on stability during prolonged storage.

Treatment with faeces suspended in saline and held at $-60\,°C$ for 14 days or for 1 day at $5\,°C$ protected chicks as well as treatment with fresh faeces;

TABLE 1

NUMBERS OF SPECIFIC BACTERIA IN 24 h CULTURES INOCULATED WITH NON-FROZEN
AND FROZEN CAECAL CONTENT OF AN ADULT COCK[a,b,c]

Bacteria	Colony forming units per ml of a culture	
	Non-frozen inoculum	Frozen inoculum
Obligate anaerobes		
Gram-positive rods		
Lactobacillus	10^8	10^7
Bifidobacterium	10^7	10^7
Eubacterium	10^6	10^4
Propionibacterium	10^6	10^5
Clostridium	10^7	10^6
Gram-negative rods		
Fusobacterium	10^6	$< 10^3$
Bacteroides	10^6	10^5
Gram-positive cocci	10^6	$< 10^5$
Facultative anaerobes		
Enterobacteria (rods)	10^9	10^7
Faecal streptococci	10^7	10^6

[a] Estimates based on isolation from 10 fold dilutions.
[b] VL medium incubated at 37 °C.
[c] Unpublished data of Schneitz, Seuna, Rizzo, Mäkelä and Nurmi.

faeces held for 6 days at room temperature did not protect.[100] Faecal material extracted from nest litter and lyophilised was also effective when resuspended.[81] Frozen cultures of faecal material also protected chicks.[69,114] A culture lyophilised in milk and subsequently added to feed, did not protect.[59]

Inocula for cultures have been preserved successfully. Some methods are: caecal content mixed 1:1 with glycerol and held at -70 °C;[97] faecal cultures frozen at an unspecified temperature after 1 to 3 subcultures;[114] lyophilised nest extracts held 12 months at 4 °C;[81] and lyophilised third serial cultures of faeces held at 4 °C (TSSF). Of three TSSF preparations, one was protective on six separate occasions during eight months but two other preparations, initially protective, lost their protective activity on storage.[68] If preserved intestinal content becomes the preferred material for inocula, then evidence must be obtained for preservation of protective activity under prolonged storage. Some changes in bacterial populations may result when media are inoculated with frozen rather than with non-frozen caecal material (Table 1).

METHODS OF CULTURING

Faecal and caecal material are, at present, the best sources of protective bacteria. Culturing these anaerobically provides sufficient material to treat a large number of chicks. In the absence of pure cultures as inocula, subculturing appears desirable. Serial subculturing dilutes intestinal material so that parasites and viruses, if present in the original inoculum, have a low probability of being present in the material given to the chicks. It may also select for more protective bacteria, but there is no published information on this point.

The following pertains mainly to cultures of intestinal material, but some pertinent information on pure cultures is included.

Media

VL medium[8] has been used widely to culture anaerobes, but there appears to have been no systematic comparison with other media to determine the most suitable medium for growing protective cultures. VL medium has been used alone, or supplemented with one or more of menadione, haemin, liver extract or sterilised poultry faecal extract.[8] In media less nutritious than VL, supplements appear necessary for growth of some pure cultures of caecal origin.[85] However, there is no substantial evidence that supplementation of VL medium produces more protective mixed cultures from caecal or faecal inocula.[74,114]

Reinforced Clostridial medium,[40] a non-selective medium designed to grow anaerobes, gave protective cultures.[104,105] MRS,[26] designed for growing lactobacilli, has been found both suitable[100] and unsuitable.[105] Trypticase soy broth grows protective cultures but not when supplemented with 500 ppm of sodium azide.[114]

Despite anaerobic culturing, several media have been found to be unsuitable for growing protective cultures: meat broth,[74] Schaedler's broth[100] and thioglycollate broth.[114]

Media have been supplemented with antibiotics and other antimicrobial agents to inhibit growth of clostridia,[96] in an attempt to explain enhanced infection by *Salmonella* in furazolidone-treated chicks[78] and in attempts to selectively inhibit non-protective bacteria while favouring growth of protective bacteria. Compounds that did not destroy the protective effect of cultures were bacitracin at 10 ppm[63] and nitrovin at 10 ppm.[75] Compounds that did destroy the protective effect were tetracycline at 20 ppm[75] and furazolidone at 100 ppm.[78] The effect of antibiotics in feed and as therapeutic agents, on the efficacy of treatment will be discussed later.

Anaerobiosis

Cultures must be grown anaerobically to be effective.[74,100] Once grown, they appear not to be damaged readily by air although methods used to detect damage may not have been sufficiently sensitive. Methods used for producing anaerobic media and culturing anaerobically have included: producing, dispensing and inoculating medium under CO_2 with or without oxygen-free N_2; Gas Pak jars; and seals of mineral oil. The CO_2 may be necessary for growth of some of the organisms essential for protection, but there has not been any attempt to verify this hypothesis for caecal or faecal cultures.

The use of media produced and maintained under anaerobic conditions may prevent the formation of compounds inhibitory to protective organisms or favour selection of fastidious anaerobes that are protective. Information on these points is lacking in the context of faecally-inoculated media.

Serial Subculturing

Faeces or caecal content have been subcultured under different conditions by several workers, but there is no standardised procedure, or even a procedure that has been shown by systematic study, to be best for producing serial subcultures.

Cultures have been serially subcultured after incubation for one, two or four days up to 19 subcultures, usually at $37\,°C$. The time for most axenic cultures of caecal origin to reach maximum populations is 1–2 days,[85] but there is no similar information for cultured caecal or faecal material. The number of subcultures and the days of incubation between subcultures may be important. Differences in protective capacity have been observed when subcultures were made under different conditions in VL medium and trypticase soy broth. The differences in protection were, however, not consistent between experiments.[114] Faecal droppings subcultured 1, 2 and 4 times at 1 day intervals in VL medium did not show decreased protective activity in replicated experiments.[68] The choice of medium may be important: the fourth serial subculture transferred at 4 day intervals in VL was more protective than similar subcultures in MRS and in Schaedler's medium.[100] Four subcultures are adequate to dilute the initial faecal inoculum to 1×10^{-10} g per treated chick and, at this dilution, viruses from the donor's faeces are unlikely to be present. The choice of $37\,°C$ for incubation may be correct, but a temperature of $40\,°C$ which is closer to that of the adult chicken may produce more protective cultures; the pertinent experiments have not been reported, although preliminary data[60] indicate no differences. There is no information on the optimum volume of culture to be transferred although this volume may affect vigour of growth, and is important in the industrial production of pure cultures for other purposes.

The number of subcultures and the volume of inoculum transferred result in a balance between the need to dilute viruses possibly present in the donor bird and the need to obtain that dilution with the fewest subcultures. As adult chickens are relatively resistant to infection by *Salmonella*, presumably their non-cultured droppings or caecal contents give the most protection against *Salmonella* and further subculturing would tend to decrease the ratio of protective bacteria. At the same time, the potential for increasing the growth of non-protective or even infectious or toxigenic pathogens may be increased. This potential hazard has not, apparently, been investigated but several reports on the use of subcultured material have not mentioned increased morbidity or mortality.

The alternative possibility that subculturing increases the ratio of protective bacteria also needs consideration.

CONSIDERATIONS ON TESTING FOR EFFICACY

Factors such as housing, methods of treatment, challenge and examination for *Salmonella*, affect the outcome and interpretation of studies of efficacy. The strain of chick appears to be unimportant as long as the chicks are not previously infected with *Salmonella* at the time of treatment. Temperature in pens or brooders must be suitable for the age of the chick or poult.

Housing
Chicks have been housed individually in boxes, in groups in wire-floored brooders, and in wire-walled or solid-walled pens on concrete floors covered with litter. Each type of housing may affect the spread of culture used for treatment and the spread of *Salmonella* used for challenge. The spread of protective culture between treated and untreated birds in the same pen has been shown;[101] although the spread of protective culture between adjacent pens may occur, more research is needed on this point. The spread of *Salmonella* between chicks housed in a single pen may be a major obstacle to precise quantitation of protection. The spread of *Salmonella* between pens, however, does not occur readily, at least for the first 8 days after challenge.[69]

Chicks housed in wire-floored brooders are not continuously exposed to *Salmonella*-contaminated faeces from infected pen mates, but those on litter are exposed. Even the location of drinking devices in wire-floored brooders, or their height above litter covered floors may make them more or less accessible to faecal contamination. None of the factors pertaining to housing have been compared to determine the effect that they have on the efficacy of treatment.

Treatment

Treatment, like housing, can affect efficacy. Thus, comparison of suspensions or cultures must be done under carefully controlled conditions. Two methods of administering treatment have been used by most workers: directly into the crop, and by addition to the drinking water.

The suspending and diluting solutions have usually been saline or peptone saline. With these diluents, chicks given from 1×10^{-3} g[48] to 0.5×10^{-1} g[62] of original tissue or gut content directly into the crop were protected. When treatment materials are to be added to drinking water in barns the dilution fluid must be chosen with great care and the concentration of protective material must be 1000 to 10 000 times higher than that used for laboratory experiments.[96] The drinking water should be warmed to avoid cold shock to the protective bacteria.

Drinking water may contain traces of chlorine or toxic metals that can be highly lethal to microorganisms.[44] The addition of 0·1 to 0·25 % powdered milk to drinking water to protect live viral vaccines is recommended.[1] Dilution of treatment culture with sterile VL medium prior to mixing with drinking water has been as effective as milk in maintaining protection.[68]

The method of administering treatment needs additional research to decrease the loss of protective bacteria that are added to drinking water in commercial barns. There is also a need for investigation of other methods for introducing protective bacteria, especially at the hatchery. Possibly aerosols could be used.[27] A varying percentage, sometimes high, of batches of chicks are already *Salmonella*-infected when shipped from the hatchery. If, in an infected hatch, a high percentage of the chicks are infected, treatment may be of little value if time between hatching and treatment is 12–24 h.[68] Treatment at the hatchery may decrease the spread of *Salmonella* in such situations.

Challenge with Salmonella

The methods of challenge in trials of efficacy are: directly into the crop; in the drinking water; seeder chicks and placement on contaminated litter. Introduction into the crop is precise and certain. If challenge is added to the drinking water, the water should be free of materials injurious to the challenge organisms, and the chicks should be deprived of water prior to challenge in order to increase thirst. Seeder birds[114] are untreated chicks infected *per os* and housed with non-infected pen mates, usually 2–3 seeders to 10–100 pen mates. Infection by seeders probably resembles most closely the spread of infection in commercial operations. Placement of chicks on litter infected by a previous flock also resembles infection that may occur in commerce.[80] The disadvantage of this method is that untreated control chicks can consume protective bacteria as well as *Salmonella* in faeces from the previous flock.[100]

Many serotypes of *Salmonella* have been used to test the efficacy of treatment in chicks and poults: *S. typhimurium, S. heidelberg, S. infantis, S. thompson, S. newington, S. arizonae.*[42,48,82,100,105] Treatment was effective against all of these, but the size of the challenge dose was important; large doses, e.g. 10^8 cells, can infect treated chicks that are resistant to 10^5 cells.

The size of the challenge dose for treated birds has been varied widely. However, only rarely has the true infectious dose for untreated birds been established to serve as a basis for comparison. One study[93] using individually housed, untreated chicks, found the ID_{50} to be 1–10 *Salmonella* cells. There has been no investigation of the true ID_{50} for treated chicks and such a study would require individually housed chicks. Hence, rigorous proof of the infectious dose for a treated chick is lacking; it may be higher than indicated by published reports.

There is a distinction to be made between the known challenge introduced into the chick by the investigator, and the continuous rechallenge of pen mates not infected directly with the known challenge but exposed to chicks that were infected. This rechallenge may be enormous as an untreated chick that becomes infected, even by 10^3 cells, will continue to excrete 10^7–10^9 *Salmonella* per gram of faeces for at least 1–2 weeks following challenge. In contrast, an adequately treated chick challenged with 10^3–10^5 *Salmonella* may destroy or totally excrete the *Salmonella* within 2–3 days without multiplication of the challenge dose during passage through the intestinal tract. Between these extremes of untreated and adequately treated chicks are the situations encountered in a normal population of animals: some do not respond fully to the treatment and are more susceptible to the challenge. Such chicks, although they may represent a small percentage of a test group, may excrete large numbers of *Salmonella*. Thus, the size of the original challenge may be negligible in relation to repeated exposure to *Salmonella* in feed and water contaminated by faeces of an infected cohort. A 10 day old chick drinks about 50 ml of water per day[1] and may consume 10^9 *Salmonella* per day if the water is heavily contaminated.[107]

Assessment of Efficacy

The efficacy of treatment will generally be evident by 10 days, perhaps less in controlled laboratory experiments. In commercial broiler flocks, assessment at 4–6 weeks may be more appropriate. An effective treatment will decrease infection. There are two main methods of determining infection in a flock, cultural and serological. Culturing is preferred because it is definitive and it need not be expensive if samples are composited. Serology, however, is useless if assessment of efficacy is made within two weeks of challenge.[113]

Treatment, as mentioned previously, is not expected to eradicate *Salmonella* infection in poultry. However, if effective, it will reduce the incidence of infected birds and in those birds that do become infected, it will reduce the numbers of *Salmonella* excreted and the duration of excretion.[113] Thus, the methods of assessing efficacy should reflect the intent of the treatment i.e. to reduce infection and to eradicate it in some flocks.

The detection of *Salmonella* depends on the sampling plan—number, weight, distribution, and frequency of collection of samples—and the sensitivity of the cultural method.[41] In some flocks, infection may indeed be absent, but for practical purposes, sampling for total absence is restricted to small flocks.

There is a need for standardised sampling plans and analytical methods in order to compare infection in treated flocks with untreated controls. Flocks may vary from 10 chicks in a wire-floored brooder to industrial scale flocks on old litter. The sampling plans will vary with the size and type of flock, but several standardised plans, each for a different type of flock, appear to be necessary.

Two criteria for infection of poultry are generally used: (a) infection of the flock and (b) the incidence of infected birds within a flock. A third criterion is useful for treated chicks: the number of *Salmonella* per gram of gut content or freshly voided faeces.

Infection in a flock is usually determined by examining litter samples collected in a specified pattern at specified intervals.[98] A single positive sample is sufficient to deem the flock infected. This system appears to be too severe to evaluate efficacy of the Nurmi concept, and less sensitive methods may be more realistic. In Sweden and Finland, a flock of broiler chickens is considered infected if any of a specified number of birds necropsied at specified intervals is found to be infected. The amount of sampling is less for broiler flocks than for breeder flocks.[2] This system of assessment will detect fewer infected flocks than examination of litter samples.

The incidence of infected birds within a flock may be assessed at 1–2 weeks of age by taking cloacal swabs, or in birds of any age by examining at necropsy the intestinal content from the ileo-caecal junction to the cloaca. The proportion of birds examined may vary from 100 % in a 10-bird flock to a statistically designed small sample in larger flocks.

The number of *Salmonella* per gram, of gut content, whether in the caecum of a sacrificed bird, in composite samples of fresh faeces or, less directly, as faecal contaminants in feed or water may be especially useful for evaluating efficacy. Such measurements indicate the numbers of *Salmonella* that may be, or have been spread from infected birds and are available to infect the non-infected pen mates. Determination of numbers will be most significant during 1–2 weeks after hatching when pen mates are most susceptible to infection.[84] Although many investigations of efficacy have involved quantitation of *Salmonella*, there has been little or no attempt to establish criteria for successful treatment based on numbers of *Salmonella* in faecal material. Such criteria appear necessary to facilitate predictions of spread of infection from infected to non-infected pen mates.

CONSIDERATIONS ON TESTING FOR SAFETY

The use of suspensions or cultures of intestinal content has created concern that pathogenic organisms might also be transferred. Suspensions have a high potential for transmitting viruses and parasites, should they be present in the donor birds. Conversely, cultures, especially subcultures, are unlikely to transmit these agents because they cannot grow in bacteriological media. Cultures may, however, transmit pathogenic bacteria which can grow in pure culture and may grow even in the competitive situation found in mixed cultures. Mycoplasma may also grow in pure cultures in VL medium, but are unlikely to grow well[79,83] or compete in mixed cultures.[60] Notwithstanding these concerns, there are no reports indicating that Nurmi treatment resulted in more deaths, more morbidity, decreased weight gains and decreased feed conversion.

Tests for absence of viruses and parasites in serially subcultured faeces is unnecessary. Tests for specific bacterial pathogens such as the following would be time-consuming and expensive: enteropathogenic *E. coli*, *Pasteurella multocida*, *Erysipelothrix rhusiopathiae*, *Staphylococcus au-*

reus, Haemophilus gallinarum, toxin forming clostridia such as *Clostridium botulinum,* and gut necrotising pathogens such as *C. perfringens*[71] and *C. colinum.*[66] An alternative to tests for all of these potential bacterial pathogens, which may or may not cause disease, would be to subject treated birds to a variety of stressors[29] that simulate adverse conditions of housing or nutrition. Failure to obtain disease in treated, stressed birds would indicate that the treatment was probably not harmful.

The possibility of transmitting antibiotic resistant bacteria to large numbers of poultry under commercial conditions may also be considered a restraint on the use of the Nurmi treatment. The transfer of such resistance by *Enterobacteriaceae* is well known especially if antibiotic-supplemented feed is used; the importance of the obligate non-sporing anaerobes has not been investigated in this context.

ANTIBIOTIC THERAPY AND NURMI TREATMENT

Antibiotics are added (where permitted) to feed at low levels, about 15–25 ppm, to stimulate growth and at high levels, 100–200 ppm, for therapy against various bacterial diseases. High levels of some antibiotics given in feed or water may suppress *Salmonella* infections in flocks, but do not eliminate them; when antibiotic treatment is terminated, *Salmonella* are excreted again. In the context of Nurmi treatment, two questions have arisen: (a) do antibiotics in the feed destroy the protective flora? and (b) can infected flocks be cleared of *Salmonella* by a combined therapy of antibiotics followed by faecal culture?

Many antibiotics added to feed do not destroy the protective effect of treatment. Chicks treated with faecal microflora and fed with antibiotic-containing feed, usually at 200 ppm (10 ppm of nitrovin), resisted subsequent challenge with *Salmonella*. The antibiotics were bacitracin, furazolidone, gallimycin, penicillin/streptomycin, chlortetracycline and tylosin.[63,75,101]

Medication of infected older birds for about one week with antibiotics to decrease the incidence and intensity of infection and subsequent introduction of faecal culture on several successive days has apparently eliminated infection[94] or decreased the number of infected chicks. In those chicks that remained infected, this combined treatment decreased the *Salmonella* per gram of faeces.[95,97] The best antibiotics for the combined therapy were mixtures of polymycin and neomycin. Such therapy may be useful when breeder flocks are infected, or when broiler flocks are infected and *Salmonella*-contaminated meat is not acceptable to regulatory authorities.

FIELD TRIALS

The widespread acceptance of Nurmi treatment will depend on its ability to decrease *Salmonella* infection in commercial poultry. Although a reduction in the proportion of *Salmonella*-infected flocks is important for breeders, a decrease in the proportion of *Salmonella*-infected birds, rather than flocks, is more important for birds raised for meat. Only a few field trials have been reported in which treated birds were compared with untreated birds. In these, the source of *Salmonella* was a contaminated environment or infection from feed or breeders; birds were not challenged with *Salmonella*. Additionally, field trials have been reported in which birds were treated, but there were no untreated controls; these cannot serve as indicators of efficacy.

Two field trials with treated and untreated commercial flocks of 5000–35 000 birds were carried out in Finland. In one trial, with 320 flocks, there was no difference in the percentage of flocks infected but there was a highly significant difference ($p < 0.001$) in the percentage (42·8 % versus 13·7 %) of chickens infected.[73] In the second trial, with 116 flocks, there was no difference in the percentage of infected flocks or chickens.[96] In these trials, the main infectious agent was *S. infantis* in the first and *S. typhimurium* var. *copenhagen* in the second. The latter may be more virulent for chickens.

In a third study,[72] flocks in five commercial operations were treated with faecal culture and carcasses were examined at slaughter. In four of the operations, about 50 % of the flocks were treated with faecal culture, but there was no attempt to correlate treatment with infection. In the fifth (Company 5 in Table 2) all flocks were treated and 207 carcasses from 34 flocks were examined. All were free of *Salmonella*. The results, given in Table 2 do not imply that treatment decreased infection as there were no untreated controls. They do, however, indicate the low incidence of *Salmonella*-contaminated carcasses in this trial.

One of the difficulties in conducting field trials is the inability to know, on the day of hatching, whether or not the flock is infected. As the percentage of naturally infected flocks is highly unpredictable, a large number of flocks must be used to obtain statistically significant results. A field trial was done with a single flock of chicks that had been hatched from eggs originating

TABLE 2

Salmonella CONTAMINATION IN COMMERCIAL CARCASSES FROM FLOCKS TREATED WITH FAECAL CULTURE WITHIN ONE DAY OF HATCHING[a,b,c]

Company	Salmonella-*positive*	
	Flocks Total (%)	*Carcasses* Total (%)
1[d]	1/21 (4·8)	5/116 (4·3)
2	7/29 (24·1)	16/160 (10)
3	3/18 (16·7)	7/129 (5·4)
4	16/43 (37·2)	33/247 (13·4)
5	0/34 (<3)	0/207 (<0·5)
Total	27/145 18·6	61/859 7·1

[a] Raevuori and Nurmi, unpublished data.[72]
[b] Results based on culture of 25 g sample of meat from area of vent.
[c] Carcasses were examined in 1978–1979. About one-half of flocks of Companies 1–4 and all flocks of Company 5 were treated.
[d] Company No. 1 processed 21 flocks and examined a total of 116 carcasses; five carcasses were found *Salmonella*-contaminated, all in one flock; the number of carcasses examined per flock is not given.

TABLE 3

EFFECT OF TREATMENT WITH FAECAL SUSPENSION ON INCIDENCE OF INFECTION IN NATURALLY CONTAMINATED CHICKS[a,b]

Treatment	Caeca positive/Caeca examined			
	Groups of pens			
	A	B	C	D
In crop	0/140	2/140	4/140	0/140
None (control)	0/140	7/140	1/140	0/140
In water	1/140	0/140	1/140	0/140
None (control)	0/140	18/140	39/140	0/140

[a] Adapted from Snoeyenbos *et al.*, 1979.[102]
[b] Chicks from a large flock were randomly assigned to 16 pens, 112 per pen, and cloacal swabs of 20 birds per pen were examined at seven weekly intervals for a total of 140 samples per pen.

from *Salmonella*-infected breeders, and the chicks were placed in pens that had previously held *Salmonella*-infected birds,[102] The pens were separated by wire walls and chicks were given treatment directly into the crop or in the water. Pens with treated birds were alternated with pens holding untreated birds. Some results, given in Table 3, indicate the paucity of useful data that may result when naturally infected chicks are used. Unfortunately, challenge of commercial flocks by introducing *Salmonella* is not acceptable to regulatory agencies despite the prevalence of infection in commercial flocks.[110]

MICROBIOLOGY OF THE CHICKEN GUT

Microbial Succession

The autochthonous flora and some associated ecological terminology are presented here to facilitate understanding of the complexity underlying the Nurmi concept. We emphasise that the microbial successions described below are a summary taken from work published over a period of 25 years during which time techniques for isolation and identification have progressed but are still incomplete. Moreover, there are numerous gaps in knowledge because of specialisation by many workers in the study of a single organ, age of chicken, genus of microorganisms or physiological activity. It is probable that, in the gastrointestinal tract of the chicken, there may be a hundred genera and several hundred species. For example, the crop of the mature chicken contains at least 14 biotypes of *Lactobacillus*, and only two of the 14 were identifiable with known species.[30] In view of the world production of poultry (about 31 000 000 tonnes[110]) and the implications of gut ecology for nutrition, feed conversion and disease control, the microbial ecology of the poultry gut merits greater attention.

Habitats in the gastrointestinal tract are colonised in a sequence dependent upon the age of the animal, the final microbial assemblage being named the climax community.[3] The forces directing the succession are autogenic (induced by the microorganisms) and allogenic (exerted by the environment, in this case the non-microbial tissues of the gut).[3] The succession leads to the establishment of the autochthonous (indigenous) flora. Criteria for the autochthonous flora of the gut have been summarised by Savage:[89] (a) they can grow anaerobically; (b) are always found in normal adults; (c) colonise particular areas of the tract; (d) colonise their habitats during succession in infant animals; (e) maintain stable population levels in climax communities in normal adults; and (f) may associate intimately with the mucosal epithelium in the area colonised.

Age plays an important role in the succession of microorganisms in the gut of chickens reared without mothers; information on succession in chicks reared by mothers is lacking, but in other animals reared by mothers, successions do occur. One can speculate that treatment of day-old chicks with adult faecal or caecal flora bypasses the normal succession and establishes immediately the climax populations of the adult donor in the chick. Or, perhaps, only a very small proportion of the adult flora is capable of establishing in the day-old chick and this proportion is also capable of preventing infection by *Salmonella*. At least some bacterial species that are not found in day-old chicks under commercial conditions can be established in germ-free chicks, but only in association with other bacteria;[57] others can be established readily.[56] This area of research remains largely unexplored in the context of succession that is related to protection.

The succession of bacteria in the gastrointestinal tract of chickens, from hatching to adulthood is summarised from several studies.[5-11,18,30,31,33,34,45,46,50,55,57,85-87] The gastrointestinal tract is a long tube with many areas of specialisation.[39] The areas pertinent to this discussion are the crop, oesophagus, proventriculus, gizzard, small intestine, and large intestine including the caecum and colon. Thus, microbial succession must be considered in the context of age of the bird, portion of the gut, and species and numbers of organisms.

The gut of the newly hatched chick is sterile, but within 12 h of hatching, and even before feeding, the content of the caecum may contain large numbers of bacteria per gram, 10^8–10^{10} faecal streptococci and coli-aerogenes[12,65] and 10^5–10^6 *Clostridium* spp.[45] The remainder of the gastrointestinal tract has, however, only low numbers of bacteria which include micrococci, clostridia and *Bacteroides*.[65] Lactobacilli are absent.[12,30,50,65]

Within 1–2 days of feeding, microbial populations increase rapidly in the duodenum, ileum and colon, but the composition remains similar to that of the chick before feeding.[65] However, lactobacilli are now present in the crop[30] at 10^6–10^7 and in the caecum[50] at 10^6–10^9 per gram.

Within 1–2 weeks, the flora in the crop, duodenum and ileum appears to have stabilised: lactobacilli predominate and persist for life at 10^5–10^8 per gram while coli-aerogenes and faecal streptococci disappear or decline to low levels.[12,30,46,65] A variety of obligate anaerobes are present, comprising 9–39% of the total flora cultured from the small intestine;

Eubacterium is the most prevalent, constituting about 20% of the total flora.[86] The caecal flora takes longer to climax.

The caecal flora, although containing 10^{10}–10^{11} bacteria per gram from about 1 day of age onward, undergoes a continuum of successions until it stabilises at about 5 weeks. For 1–2 days, coli-aerogenes and faecal streptococci predominate. At 2–3 days[50] lactobacilli reach populations of 10^8–10^9 and persist at these levels for a few weeks in some chickens, or become only a minor component of the caecal flora in others.[86] As the chicken ages, all three of the above groups decrease as percentages of the total caecal population and by two weeks they comprise less than 1% of the caecal flora; the remainder are obligately anaerobic. About half of the total number of obligate anaerobes observed microscopically in caeca of two week old chickens have been enumerated in culture media. Those cultured are: *Eubacterium*, 60·6%; anaerobic cocci, 14·2%; *Bacteroides*, 12·8%; *Fusobacterium*, 6·2%; *Gemmiger*, 3·4% and *Clostridium* (found in only 1 of 6 caeca), 2·1%.[86]

At 5–12 weeks of age the caecal bacteria consist primarily of obligate anaerobes: *Bacteroidaceae*, *Propionibacterium*, *Eubacterium*, *Peptostreptococcus*, Gram-negative pleomorphic cocci and *Clostridium*.[7,10,85] Some of the morphological types seen in caecal material by microscopy are apparently not culturable with methods presently available. The relative proportions of the various species recognised in the chicken gut may be influenced by breed, diet, geography and methods of isolation.[87]

Against this background, we will discuss the use of pure cultures as inhibitors of infection by *Salmonella*.

Pure Cultures as Protective Bacteria

The loss of protection against *Salmonella* in mice after treatment with antibiotics has been mentioned previously. The protection was restored partially by dosing the mice with some strains of *Bacteroides* and restored further when *Escherichia coli* was given concomitantly; *E. coli* by itself conferred no protection. Thus, more than one species is required to protect mice.[54] It is important, in all experiments with pure cultures, to recognise that strains of *Bacteroides* that were protective in mice when freshly isolated, lost their protective ability on repeated subculture.[54] Mutation, with loss of ability to form a glycocalyx, may have caused the decreased protection by *Bacteroides* and is a possible explanation for observations reported below with *Streptococcus faecalis*.

There is relatively little information on the protective activity of pure cultures, isolated from the gut of adult chickens, against infection by

Salmonella in newly hatched chicks. Only two cultures have been reported to prevent infection, both in Australia.[105,106] One of these, *Strep. faecalis*, protected chicks against *S. typhimurium* ($p < 0.01$), and *S. newington;* protection against *S. anatum* was equivocal. In view of the natural, rapid spread and high populations of *Strep. faecalis* in day-old chicks, which nevertheless are still susceptible to infection by 1–10 cells of *Salmonella*, *Strep. faecalis* appears to be unlikely as a protective species. The culture used in Australia was investigated independently in the Netherlands[37] and in Canada,[68] Neither study demonstrated protective activity. Additionally, two collections of *Strep. faecalis*, var. *faecalis* and *Strep. faecalis* var. *liquefaciens* freshly isolated from chicken faeces were used in the Canadian studies; both serotypes 8 and 9 were included. The Australian studies[105,106] used *Strep. faecalis* var. *faecalis*, serotype 8. All identification and serotyping was done or confirmed by Dr J. C. Huang, Ottawa, Canada. The second culture found to protect chicks in Australia against *Salmonella* was *Bacteroides fragilis*. Unfortunately, it lost viability in storage.[105]

Some pure cultures have been found to be partially protective: an unnamed *Clostridium*[82] and unnamed bacterial isolates (genera not stated).[102]

Pure cultures that have not shown protection are *Bacteroides hypermegas* and *Bifidobacterium* spp.,[10] *E. coli*[42,100] and the following mixtures of pure cultures: strains of lactobacilli; lactobacilli with *Bacteroides vulgatus* and *Bifidobacterium* sp.; lactobacilli with two strains of *Bifidobacterium* sp. plus an unidentified anaerobe, possibly *Eubacterium* sp.[9] Even worse, *Salmonella* infection in chicks may be exacerbated if treatment with large numbers of only a few types of bacteria is given prior to challenge. In this context, cultures that prevented *E. coli* from growing in the gut resulted in a heavier growth of *Salmonella*. The causes of these unexpected observations are unknown.[9]

Mixtures of pure cultures that can inhibit *Salmonella* infection of newly hatched poultry may be discovered; at present, there is intense interest in a few laboratories. Recently a mixture of pure cultures given to newly hatched chicks has been shown to reduce the incidence and severity of infection by *S. typhimurium* given one day later.[117] The discovery of such mixtures would facilitate employment of the Nurmi concept in those countries that still have concern about the safety of faecal cultures. However, there is a pressing need to test faeces or faecal cultures in carefully designed field trials using commercial flocks in several countries. If such field trials support the findings of the successful field trial conducted in Finland, then the search for mixtures of pure cultures that are protective will be more justified.

CONCLUSIONS

The widespread use of treatment based on the Nurmi concept requires that the treatment will not harm poultry and that it will decrease infection under commercial conditions. The most desirable point of application would be at the hatchery, but if treatment is given in the drinking water in the barn, then diluents are required to preserve the protective bacteria until the treatment is consumed. If carefully controlled trials in several countries, using caecal or faecal cultures, show substantial reduction of infection, then the search for pure cultures may be carried on with considerable assurance of success.

REFERENCES

1. Agriculture Canada, *Broiler Raising in Canada*, Publication 1509, 1976, Information Division, Canada Department of Agriculture, Ottawa.
2. AKERMAN, H., and HUGOSON, G., *Salmonella* control: poultry, *LBS*, 1975, **27** Vb 20, pp. 1–7 (Sweden). Original in Swedish.
3. ALEXANDER, M., *Microbial Ecology*, 1971, John Wiley and Sons, Inc., New York.
4. ALMLOF, J., *Proc. Intern. Symp. on* Salmonella *and Prospects for Control*, ed. Barnum, D. A., 1977, Univ. Guelph, Canada.
5. BARNES, E. M., *Amer. J. Clin. Nutr.*, 1972, **25**, 1475.
6. BARNES, E. M. and IMPEY, C. S., *J. Appl. Bact.*, 1968, **31**, 530.
7. BARNES, E. M. and IMPEY, C. S., *Br. Poultry Sci.*, 1970, **11**, 467.
8. BARNES, E. M. and IMPEY, C. S., *Isolation of Anaerobes*, SAB Technical Series No. 5, eds. Shapton, D. A., and Board, R. G., 1971, Academic Press, London.
9. BARNES, E. M., and IMPEY, C. S., *The Vet. Rec.*, 1980, **106**, 61.
10. BARNES, E. M., IMPEY, C. S., and STEVENS, B. J. H., *J. Hyg. Camb.*, 1979, **82**, 263.
11. BARNES, E. M., MEAD, G. C., BARNUM, D. A., and HARRY, E. G., *Br. Poultry Sci.*, 1972, **13**, 311.
12. BARNES, E. M., MEAD, G. C., IMPEY, C. S., and ADAMS, B. W., *Br. Poultry Sci.*, 1978, **19**, 713.
13. BARNUM, D. Unpublished data.
14. BOHNHOFF, M., DRAKE, B. L., and MILLER, C. P., *Proc. Soc. Exper. Biol. and Med.*, 1954, **86**, 132.
15. BOHNHOFF, M., MILLER, C. P., and MARTIN, W. R., *J. Exper. Med.*, 1964, **120**, 805.
16. BOHNHOFF, M., MILLER, C. P., and MARTIN, W. R., *J. Exper. Med.*, 1964, **120**, 817.
17. BOWMAN, P. J., CUMMING, R. B. and LLOYD, A. B., *Proc. First Australasian Poultry and Stock Feed Convention*, 1976, Melbourne, Australia.
18. BROOKER, B. E., and FULLER, R., *J. Ultrastructure Res.*, 1975, **52**, 21.
19. BROWNELL, J. R., SADLER, W. W., and FANELLI, M. J., *Avian Dis.*, 1970, **14**, 106.

20. CARTER, P. B., *Microbiology*, ed. Schlessinger, D., 1975, Amer. Soc. for Microbiol., Washington, DC.
21. CHENG, K.-J., MCCOWAN, R. P., and COSTERTON, J. W., *Amer. J. Clin. Nutr.*, 1979, **32**, 139.
22. CLARK, G. M., KAUFMANN, A. F. and GANGAROSA, E. J., *The Lancet*, Sept. 1973, **1**, 490.
23. COCKRILL, W. R. *Poultry Disease and World Economy*, eds. Gordon, R. F., and Freeman, B. M., 1971, British Poultry Science, Edinburgh.
24. COSTERTON, J. W., GEESEY, G. G. and CHENG, K.-J., *Scientific American*, 1978, **238**, 86.
25. COX, N. A., DAVIS, B. H., WATTS, A. B., and COLMER, A. R., *Poultry Sci.*, 1973, **52**, 661.
26. DEMAN, J. C., ROGOSA, M., and SHARPE, M. E., *J. Appl. Bact.*, 1960, **23**, 130.
27. FAGERBERG, D. J., AVENS, J. S., and QUARLES, C. L., *Proc. and Abstr. 15th World's Poultry Congress and Exposition*, 1974, New Orleans, USA.
28. FANELLI, M. J., SADLER, W. W., FRANTI, C. E., and BROWNELL, J. R., *Avian Dis.*, 1971, **15**, 366.
29. FRASER, D., RITCHIE, J. S. D., and FRASER, A. F., *Br. Vet. J.*, 1975, **131**, 653.
30. FULLER, R., *J. Appl. Bact.*, 1973, **36**, 131.
31. FULLER, R., *J. Appl. Bact.*, 1978, **45**, 389.
32. FULLER, R., and BROOKER, B. E., *Amer. J. Clin. Nutr.*, 1974, **27**, 1305.
33. FULLER, R., COATES, M. E., and HARRISON, G. F., *J. Appl. Bact.*, 1979, **46**, 335.
34. FULLER, R., and TURVEY, A., *J. Appl. Bact.*, 1971, **34**, 617.
35. Gordon Research Conferences, *Science*, 1979, **203**, 1141.
36. GOREN, E., *Proc. 21st World Vet. Congr.*, 1979, Moscow, USSR.
37. GOREN, E. Unpublished data.
38. HAUSCHILD, A. H. W. and BRYAN, F., *J. Food Protect.*, 1980, **43**, 435.
39. HILL, K. J., *Physiology and Biochemistry of the Domestic Fowl*, Vol. 1., eds. Bell, D. J., and Freeman, B. M., 1971, Academic Press, London and New York.
40. HIRSCH, A., and GRINSTED, E., *J. Dairy Res.*, 1954, **21**, 101.
41. ICMSF, *Microorganisms in Foods. 2. Sampling Plans for Microbiological Analysis: Principles and Specific Applications*, 1974, University of Toronto Press, Toronto, Canada, 213 pp.
42. IDZIAK, E. S., and CALDWELL, M., *Proc. Intern. Symp. on Salmonella and Prospects for Control*, ed. Barnum, D. A., 1977, Univ. Guelph, Canada.
43. JACKSON, C. A. W., LINDSAY, M. J., and SHIEL, F., *Australian Vet. J.*, 1971, **47**, 485.
44. JORDAN, F. T. W., and NASSAR, T. J., *Avian Pathol.*, 1973, **2**, 91.
45. LEV, M. and BRIGGS, C. A. E., *J. Appl. Bact.*, 1956, **19**, 36.
46. LEV, M., and BRIGGS, C. A. E., *J. Appl. Bact.*, 1956, **19**, 224.
47. LLOYD, A. B., CUMMING, R. B., and KENT, R. D., *Proc. 1974 Australasian Poultry Science Convention*, 1974, Hobart, Australia.
48. LLOYD, A. B., CUMMING, R. B., and KENT, R. D., *Australian Vet. J.*, 1977, **53**, 82.
49. MARTHEDRAL, H. E., *Proc. Intern. Symp. on Salmonella and Prospects for Control*, ed. Barnum, D. A., 1977, Univ. Guelph, Canada.

50. MEAD, G. C., and ADAMS, B. W., *Br. Poultry Sci.*, 1975, **16**, 169.
51. MEYNELL, G. G., *Br. J. Exper. Pathol.*, 1963, **44**, 209.
52. MEYNELL, G. G., and SUBBAIAH, T. V., *Br. J. Exper. Pathol.*, 1963, **44**, 187.
53. MILLER, C. P., and BOHNHOFF, M., *J. Infect. Dis.*, 1962, **111**, 107.
54. MILLER, C. P., and BOHNHOFF, M., *J. Infect. Dis.*, 1963, **113**, 59.
55. MORISHITA, Y., and MITSUOKA, T., *Jap. J. Microbiol.*, 1976, **20**, 197.
56. MORISHITA, Y., MITSUOKA, T., KANEUCHI, C., YAMAMOTO, S., and OGATA, M. *Jap. J. Microbiol.*, 1971, **15**, 531.
57. MORISHITA, Y., MITSUOKA, T., KANEUCHI, C., YAMAMOTO, T., YAMAMOTO, S., and OGATA, M., *Jap. J. Microbiol.*, 1972, **16**, 27.
58. MORRIS, G. K., BIRCH, L. M., GALTON, M. M., and WELLS, J. C., *Amer. J. Vet. Res.*, 1969, **30**, 1413.
59. NOTERMANS, S., OOSTEROM, J., and VAN SCHOTHORST, M., *Fleischwirtschaft*, 1979, **59**, 733.
60. NURMI, E. Unpublished data.
61. NURMI, E., and PIVNICK, H. Unpublished calculations.
62. NURMI, E., and RANTALA, M., *Nature*, 1973, **241**, 210.
63. NURMI, E. and RANTALA, M., *Res. Vet. Sci.*, 1971, **17**, 24.
64. NURMI, E., SEUNA, E., and RAEVUORI, M., *Proc. Intern. Symp. on* Salmonella *and Prospects for Control*, ed. Barnum, D. A., 1977, Univ. Guelph, Canada.
65. OCHI, Y., MITSUOKA, T., and SEGA, T., *Zentrablatt für Bakt. Parasit. Infekt. Hyg. 1. Orig.*, 1964, **193**, 80.
66. ONONIWU, J. C., PRESCOTT, J. F., CARLSON, H. C., and JULIAN, R. J., *Can. Vet. J.*, 1978, **19**, 226.
67. PIVNICK, H., *Proc. 82nd Annual Meeting*, US Animal Health Assoc., 1978, Buffalo, NY, USA.
68. PIVNICK, H., and BLANCHFIELD, B., Unpublished data.
69. PIVNICK, H., BLANCHFIELD, B., D'AOUST, J.-Y., RIGBY, C., and PETTIT, J., Unpublished data.
70. PIVNICK, H., HANDZEL, S., and LIOR, H., *Proc. Intern. Symp. on* Salmonella *and Prospects for Control*, ed. Barnum, D. A., 1977, Univ. Guelph, Canada.
71. PRESCOTT, J. F., SIVENDRA, R., and BARNUM, D. A., *Can. Vet. J.*, 1978, **19**, 181.
72. RAEVUORI, M., and NURMI, E., Unpublished data.
73. RAEVUORI, M., SEUNA, E., and NURMI, E., *Acta Vet. Scand.*, 1978, **19**, 317.
74. RANTALA, M., *Acta Path. Microbiol. Scand. Section B.*, 1974, **82**, 75.
75. RANTALA, M., *Br. Poultry Sci.*, 1974, **15**, 299.
76. RANTALA, M., The influence of the intestinal flora and some antimicrobial drugs on the colonization of *Salmonella infantis* in the intestine of the chicken, 1976, Ph.D. Thesis, College of Veterinary Medicine, Helsinki.
77. RANTALA, M., and NURMI, E., *Br. Poultry Sci.*, 1973, **14**, 627.
78. RANTALA, M., and NURMI, E., *J. Hyg. Camb.*, 1974, **72**, 349.
79. RAZIN, S., and TULLY, J. G., *J. Bacteriol.*, 1972, **102**, 306.
80. RIGBY, C. E., and PETTIT, J. R., *Avian Dis.*, 1979, **23**, 442.
81. RIGBY, C., and PETTIT, J. R., *Avian Dis.*, **24**, 604.
82. RIGBY, C., PETTIT, J., and ROBERTSON, A., *Proc. Intern. Symp. on* Salmonella *and Prospects for Control*, ed. Barnum, D. A., 1977, Univ. Guelph, Canada.

83. RODWELL, A. W., and MITCHELL, A. *The Mycoplasmas I. Cell Biology*, eds. Barile, M. F., and RAZIN, S. 1979, Academic Press, New York.
84. SADLER, W. W., BROWNELL, J. R., and FANELLI, M. J., *Avian Dis.*, 1969, **13**, 793.
85. SALANITRO, J. P., BLAKE, I. G., and MUIRHEAD, P. A., *Appl. Microbiol.*, 1974, **28**, 439.
86. SALANITRO, J. P., BLAKE, I. G., MUIRHEAD, P. A., MAGLIO, M., and GOODMAN, J. R., *Appl. and Envir. Microbiol.*, 1978, **35**, 782.
87. SALANITRO, J. P., FAIRCHILDS, I. G., and ZGORNICKI, Y. D., *Appl. Microbiol.*, 1974, **27**, 678.
88. SAVAGE, D. C., *Microbiology*, ed. Schlessinger, D., 1975, Amer. Soc. for Microbiol., Washington, DC, USA.
89. SAVAGE, D. C., *Ann. Rev. Microbiol.*, 1975, **31**, 107.
90. SAVAGE, D. C., *Microbial Ecology of the Gut*, eds. Clarke, R. T. J., and Bauchop, T., 1977, Academic Press, London and New York.
91. SAVAGE, D. C., and BLUMERSHINE, R. V. H., *Infect. and Immun.*, 1974, **10**, 240.
92. SAVAGE, D. C., *Amer. J. Clin. Nutr.*, 1979, **32**, 113.
93. SEUNA, E., *Avian Dis.*, 1979, **23**, 392.
94. SEUNA, E., ANDERSSON, P., and NURMI, E., *Proc. 13th Nordic Vet. Congr.*, 1978, Turku, Finland.
95. SEUNA, E., and NURMI, E., *Poultry Sci.*, 1979, **58**, 1171.
96. SEUNA, E., RAEVUORI, M., and NURMI, E., *Br. Poultry Sci.*, 1978, **19**, 309.
97. SEUNA, E., SCHNEITZ, C., NURMI, E., and MÄKELÄ, P. H., *Poultry Sci.*, 1980, **59**, 1187.
98. SNOEYENBOS, G. H., *Avian Dis.*, 1971, **15**, 28.
99. SNOEYENBOS, G. H., CARLSON, V. L., SMYSER, C. F., and OLESIUK, O. M., *Avian Dis.*, 1969, **13**, 72.
100. SNOEYENBOS, G. H., WEINACK, O. M., and SMYSER, C. F., *Avian Dis.*, 1978, **22**, 273.
101. SNOEYENBOS, G. H., WEINACK, O. M., and SMYSER, C. F., *Avian Dis.*, 1979, **23**, 904.
102. SNOEYENBOS, G. H., WEINACK, O. M., and SMYSER, C. F., *Proc. 21st World Vet. Congr.*, 1979, Moscow, USSR, July, 1 7.
103. SNOEYENBOS, G. H., WEINACK, O. M., SMYSER, C. F., WESTON, C. R., and SMITH, J. H., *52nd Northeastern Conference on Avian Diseases*, 1980, Cornell Univ., Ithaca, NY, USA.
104. SOERJADI, A. S., *Proc. of the Second Australasian Poultry and Stock Feed Convention*, 1978, Sydney, Australia.
105. SOERJADI, A. S., Prevention of enteric salmonellosis in young chickens by early establishment of the normal intestinal microflora, 1979, Ph.D. Thesis, University of New England, Armidale, NSW, Australia, 187 pp.
106. SOERJADI, A. S., LLOYD, A. B., and CUMMING, R. B., *Australian Vet. J.*, 1978, **54**, 549.
107. STERSKY, A., THACKER, C., BLANCHFIELD, B., and PIVNICK, H. Unpublished data.
108. STEVENS, A. J., *Poultry Disease and World Economy*, eds. Gordon, R. F., and Freeman, B. M. 1971, British Poultry Science, Edinburgh.

109. TAKEUCHI, A., *Microbiology*, ed. Schlessinger, D., 1975, Amer. Soc. for Microbiol., Washington, DC.
110. TODD, E., *J. Food Protect.*, 1980, **43**, 129.
111. TURNBULL, P. C. B., and RICHMOND, J. E., *Br. J. Exper. Pathol.*, 1978, **59**, 64.
112. VASENIUS, H., and JAHKOLA, M., Report of the Second Congress, International Society for Animal Hygiene, 1976, Zagreb.
113. WEINACK, O. M., SMYSER, C. F., and SNOEYENBOS, G. H., *Avian Dis.*, 1979, **23**, 179.
114. WEINACK, O. M., SNOEYENBOS, G. H., and SMYSER, C. F., *Avian Dis.*, 1979, **23**, 1019.
115. WILSON, S. T., *Proc. National Salmonellosis Seminar*, 1978, US Department of Agriculture, Washington, DC, USA.
116. ZECHA, B. C., McCAPES, R. H., DUNGAN, W. M., HOLTE, R. J., WORCESTER, W. W., and WILLIAMS, J. E., *Avian Dis.*, 1977, **22**, 141.
117. BARNES, E. M., IMPEY, C. S., and COOPER, D. M., *Amer. J. Clin. Nutr.*, 1980, **33**, 2426.

Chapter 3

THE BACTERIOLOGY OF FISH HANDLING AND PROCESSING

G. Hobbs and W. Hodgkiss

*Ministry of Agriculture, Fisheries and Food,
Aberdeen, UK*

SUMMARY

Spoilage and minimisation of health risks are the two main concerns of bacteriological studies of fish.

With the exception of heat sterilised and frozen products, growth of some bacteria can occur in all preserved fish products even when properly stored. Where faults occur in handling and processing bacteriological problems are more serious and can arise with all preserved products.

During the past 15–20 years the emphasis of bacteriological studies has moved away from spoilage problems towards questions of hygiene and food poisoning. This is reflected partly in the increased interest in and application of bacteriological specifications.

Studies on the spoilage bacteria have resulted in some clarification of their taxonomy with changes in nomenclature, and associations of certain bacteria with particular spoilage changes are beginning to emerge.

Although most food poisoning risks can be associated with fish, three are especially so. Poisoning caused by Vibrio parahaemolyticus, Clostridium botulinum *and scombrotoxin have been discussed in some detail.*

INTRODUCTION

The bacteriological changes that occur during the handling and processing of fish and fishery products depend upon many factors. Fish and shellfish are cold-blooded animals and the bacterial flora of the live fish will therefore be greatly influenced by the environment. The gross chemical composition of fish species can vary widely: 'non-fatty' fish such as cod and haddock commonly have 20% protein, 1% fat, 1% carbohydrate and 80% water whereas fatty fish such as herring or mackerel can have 15–20% protein, 3–25% fat, 1% carbohydrate and 70% water. These components also vary markedly, depending on the season and sexual maturity of the individual fish. A comprehensive review of the biological variations of fish has been presented by Love.[1] Once fish is caught subsequent changes in the bacterial flora will depend upon the conditions imposed by the processing procedures and the flora introduced during these. Changes in the bacterial flora and the degree to which these bacteria are able to grow will determine the spoilage pattern of the particular product and the degree of public health hazard present when the fish is consumed.

Previous reviews of the subject[2–5] have surveyed the work carried out from the pioneering investigations up to the time when current methods of preservation were well established in the developed countries. When efficient freezing, cold-storage and cold-chain distribution are used, most bacterial spoilage problems are obviated if the raw material has been properly handled. However, chill-storage and distribution as well as chill-display are not without bacterial spoilage problems.

Current interests reflect a change in emphasis in research activities over the past twenty years towards investigations into possible health hazards and the application of bacteriological specifications.

One of the factors which has influenced these trends is the increase in international trade in fishery products, particularly in tropical shellfish, others are the introduction of novel raw materials such as minced fish, the exploitation of different species of fish and the increasing popularity of convenience foods. Changes in international fishing limits have also forced changes in the balance of different fish species available for processing.

SPOILAGE

Spoilage of Fish and Shellfish

The early work on the role of bacteria in the spoilage of fishery products was reviewed by Shewan.[2,3] These reviews also presented details of those biochemical changes which occur in fish flesh as a result of bacterial activity with special reference to the rate of formation of compounds which might be used as objective indices of freshness.[6,7] The bacterial populations present on fresh fish and the changes occurring during chill-storage (usually in ice) were well documented. Similarly, the changes in the flora as a result of the handling and processing methods then in use in the fishing industry had also been investigated. The effects on the flora of salting and smoke curing and special forms of spoilage such as that caused by extreme halophilic bacteria in dry-salted products were also well recognised. In refrigerated storage of 'wet' fish, it was clear that Gram-negative psychrophilic or psychrotrophic organisms[8] were the important agents of spoilage. However, the identification of these Gram-negative bacteria isolated from fresh and spoiling fish was a difficult and uncertain procedure. From a taxonomic point of view they could at best be placed in genera and even then some of these genera were not well defined. Interest, therefore, became centred upon better characterisation of the different types of bacteria found on fishery products.[9] Indeed, this was an interest pursued by bacteriologists involved in the study of the spoilage of all forms of chill-stored protein foods.[10]

The initial difficulties of identifying and classifying these psychrophilic organisms had been due, not only to the lack of determinative criteria and suitable laboratory tests, but also to the confused state of the nomenclature and taxonomy of them at that time.[11] Since then the constantly developing methodology of bacterial identification and changing views on nomenclature and taxonomy have resulted in changes in the classification of many of these organisms.[12] Thus, organisms which, in early publications (say before 1957),[13] were assigned to the genus *Achromobacter* were a very mixed collection. The genus *Achromobacter* as then defined in the 6th Edition of Bergey's Manual[14] included (1) non-motile rods which are now called either *Moraxella*[15] or *Acinetobacter*[16] depending upon their cytochrome oxidase activity, (2) motile peritrichous rods which would now be assigned to the genus *Alcaligenes*[17,18] and (3) motile, non-fluorescent rods with polar flagella now identified, depending upon the moles % GC in their DNA, as *Pseudomonas* spp. and *Alteromonas* spp.[19,20,21] It is necessary to consider these changes when referring to earlier work even up to the review by Shewan and Hobbs.[5] In a review of the bacterial flora of freshly caught teleost and elasmobranch fish Horsley[22] comments upon some of these changes whilst Shewan[4,23] and Liston[24] provide a more comprehensive view of the present situation. Unless details are given of the methods and criteria which were used to identify the organisms, it is, however, very difficult to relate early data on flora studies to present-day concepts of the genera concerned.

Most of the literature on the bacterial flora of fish refers to marine species, usually because these have been widely exploited commercially and have warranted more intensive study than the freshwater species.[4,22,23,24]

These reviews also show that there is, for the same reason, a relative paucity of information on tropical fish.

A few studies have been reported and these indicate that, as with marine fish, the flora of freshwater fish reflects that of the environment and feed.[25,26,27] One study indicates that carp has a longer shelf life than marine fish and this correlates with slower bacterial growth in this freshwater fish.[28] The shelf life of freshwater fish is reduced by physical damage and when the fish is feeding,[27] again the latter is a situation similar to one which applies to marine fish.

The pattern of spoilage of refrigerated 'wet' fish was summarised by Hodgkiss[29] who described the spoilage flora of iced cod as an example but this simplified pattern is not true of all fishery products (*vide infra*). Freshly caught fish carry a mixed, commensal bacterial flora consisting, in the main, of organisms which belong to the Gram-negative genera *Pseudomonas, Alteromonas, Acinetobacter, Vibrio, Flavobacterium* and *Cytophaga* and Gram-positive organisms of the coryneform group and the genus *Micrococcus*. The proportion of these different types in the commensal flora varies quite considerably as reference to the review by Horsley shows.[22] One reason for this was pointed out by Shewan and Hobbs,[5] the flora on the outer surfaces and in the intestine of fish is largely determined by the flora of the environment and the feed respectively. Thus, even in temperate and sub-Arctic waters large variations occur through Gram-negative organisms (*Pseudomonas–Moraxella–Acinetobacter*) generally predominate.[30] On fish caught in Australian coastal waters a Gram-positive flora is predominant[31] whilst investigations in other areas have shown that *Vibrio* spp. form an important proportion of both the commensal and spoilage flora of certain fish.[32–35]

During chill-storage the psychrophilic strains multiply rapidly, invade the tissues and eventually cause putrefaction. Organisms of the *Pseudomonas–Alteromonas* and *Moraxella–Acinetobacter* groups increase in numbers more quickly than the other organisms, and in the later stages of spoilage comprise 80% of the total flora. Putrefaction proceeds rapidly even at chill temperatures once the bacterial load attains or exceeds 10^6 organisms per gram. The rate at which putrefaction proceeds depends upon the rate of growth of these organisms and this is largely governed by two factors—the temperature and the number of organisms present.

A proposal has been made that it is possible to use a single relative rate function to describe all types of fish spoilage reactions whether bacteriological, chemical, physical or organoleptic.[36] It is claimed that the relative spoilage rate is sufficiently well described, by the mathematical function proposed, to be useful over a temperature range of 0–15 °C. The errors in estimating overall spoilage are large and the authors have shown empirically that the relationships they propose hold with a variety of published spoilage data. The theoretical basis for these relationships, however, is questionable. The work may well have practical value, nevertheless, since it could permit the calibration of temperature function integrators[36] which could then be used to monitor storage and distribution of fish and other foods. In practice proper chilling and good hygiene are the two most important factors contributing to the prevention of spoilage.

Interest in identifying the types of organisms on fresh and spoiling fish has usually been with a view to detecting the 'active spoilers' in the flora. Before proceeding to deal with this particular aspect of fish spoilage, changes other than those already discussed in relation to flora analyses ought to be mentioned. During the past twenty years many refinements and changes in the techniques of estimating bacterial populations and detecting bacterial growth have occurred. Most of the early estimations were carried out using the pour plate technique, with 1·0 ml amounts of serial dilutions of flesh or skin homogenates, crudely prepared by grinding or shaking with sand. It is now recognised that exposure to the temperature of molten agar in the pour plate technique is sufficient to kill a proportion of psychrophilic organisms.[8] It is preferable, therefore, to use surface inocula of 0·1 ml amounts of appropriate dilutions dispensed by the use of semi-automatic pipettes.[37,38] In the preparation of initial samples the tendency has been toward the use of purpose-made power-driven homogenisers or the Stomacher.[39] The type of diluent, culture media[40,41] and the temperature of incubation[42] markedly affect the numbers and types of organisms isolated. If it is accepted that counts and flora analyses are intended to determine those organisms which are actively growing in a particular environment then the media and incubation conditions should approximate to those in the sample under test. Other advances[43] such as automatic plate pouring machines,[44] the application of Spiral-plate makers[45,46] automatic electric colony counters[47] are currently reducing the tedium of the detection and enumeration of bacteria.

Many attempts have been made to formulate methods to either assess the number of 'active spoilers'[48] or to use the detection of and measurement of metabolic products of these organisms in foods as indices of quality.[6,7]

The role of *Alteromonas* (*Pseudomonas*) *putrefaciens* in fish spoilage was re-investigated by Levin[49] using a peptone–iron agar to assess the presence of H_2S producing organisms. Sadovski and Levin[50] studied the nuclease activity of fish spoilage bacteria, fish pathogens and related species and proposed that DNAase activity could be used as a criterion for the identification of active spoilage organisms. Levin and his coworkers[51] continued their studies into the importance of pseudomonads in spoilage flora. Using gas chromatography and mass spectrometry they also showed that phenylethyl alcohol and phenol detected in refrigerated haddock fillets were metabolic by-products of Moraxella-like organisms.[52,53]

Farber and his colleagues[54] using sterile press muscle juice of English sole showed that *Pseudomonas* spp. (belonging to Groups I, II and IV of Shewan, Hobbs and Hodgkiss[9]), *Aeromonas* spp. and *Vibrio* spp. were the most active spoilers at 5 °C. Shaw and Shewan[55] using the press muscle juice technique of Farber and ethylene oxide-sterilised muscle blocks[56] considered that *Pseudomonas* spp. of Groups II, III and IV were particularly active in the spoilage of cod and haddock stored at temperatures between $-6°$ and $+6°$C. Herbert, Ellis and Shewan[57] in biochemical studies of iced cod showed that hydrogen sulphide, methyl mercaptan and dimethylsulphide were responsible for the 'sulphidy' odours associated with stale iced cod. This work was followed by studies which showed that thirteen strains of *Pseudomonas* and *Alteromonas* were active in the production of H_2S and methyl mercaptan and six of these strains also produced dimethylsulphide. There was no evidence that autolytic enzymes were involved in the production of volatile sulphur compounds.[58] In a series of studies on sterile fish muscle homogenates inoculated with a wide range of 'active spoilers' (*Pseudomonas perolens*, *Ps. putrefaciens*, *Ps. fluorescens*, *Ps. fragi* and *Achromobacter* (*Moraxella*) spp.) Miller and his colleagues[59−62] using gas–liquid chromatography and mass spectrometry were able to detect a wide range of metabolic by-products. These authors considered that certain compounds, detectable by their methods, were responsible for specific off odours; e.g. *Pseudomonas fragi* produces 'fruity' off odours which they attributed to the synergistic flavour interactions of ethyl esters.

In studies of the spoilage of refrigerated haddock fillets, Laycock and Regier[63] followed the changes in the bacterial flora in relation to the concentration of trimethylamine in the tissues and concluded that strains of *Alteromonas putrefaciens* were mainly responsible for the production of this compound.

Easter *et al.*[64] have shown that some spoilage organisms, including *Alteromonas putrefaciens*, are able to utilise trimethylamine oxide (TMAO) as a hydrogen acceptor as an alternative to oxygen. This enables non-fermentative organisms to grow under micro-acrophilic or anaerobic conditions in tissues. This is offered as one reason for the fact that fish tissues spoil more quickly than those of other animal flesh foods which contain less TMAO. Van Spreckens[65] inoculated sterilised fish muscle with pure cultures of bacteria and concluded that strains of *Alteromonas* (*Pseudomonas*) *putrefaciens* and organisms of a similar, 'non-defined' group were the main spoilage agents in chilled fish.

Shewan and Murray[66] reviewed the role of psychrophilic organisms in fish spoilage and the organoleptic changes produced in sterile muscle inoculated with pure cultures. One of the aspects which they discussed was the differing spoilage rates of various species of fish stored in ice.

There is good evidence that both fish and shellfish caught in tropical waters have a longer shelf life when stored in ice than those caught in temperate and sub-Arctic waters.[23,67] One explanation for this long shelf life of tropical fish compared with those from temperate waters is that a psychrophilic flora has to develop in the former. The difference between the shelf life of flat fish such as plaice and sole and that of cod could be due to the anti-bacterial activity of the surface slime. Plaice slime has been shown to possess a powerful lysozyme which is not present in the slime of cod.[68]

In recent years there has been appreciable interest in the microbiology of shellfish both from the point of view of the normal and spoilage flora and the general hygiene of processing. Much of this work was reviewed by Cann.[67,69]

The bacterial flora of freshly caught shellfish is similar to that of other fish. It is varied in detail but the major groups present are *Micrococcus*, coryneforms, *Moraxella*, *Acinetobacter* and *Pseudomonas* with smaller numbers of *Flavobacterium*, *Cytophaga* and *Bacillus* species. This is borne out by the work of Lee and Pfeifer on Dungeness crab (*Cancer magister*)[70] and on Pacific shrimp (*Pandalus jordani*).[71] An exception was reported by Koburger *et al.*[72] who found that the flora of freshly caught rock shrimp (*Sicyonia brevirostris*) was predominantly *Flavobacterium* and *Cytophaga* species.

A number of studies of pond reared shrimp have been carried out and these have shown that the flora is essentially the same as that of other shrimp, indicating that pond rearing presents no peculiar problems in this respect.[73-78]

There are fewer studies on other shellfish but the indications are that the fresh and spoilage flora is essentially the same. Cox and Lovell[79] demonstrated a *Pseudomonas/Achromobacter* spoilage flora in freshwater crayfish where the active 'spoilers' were mainly *Pseudomonas* sp.

Webb and Thomas[80] found similar results with scallop meats and Vasconcelos and Lee[81] demonstrated a typical Gram-negative flora in Pacific oysters (*Crassostrea gigas*). They also showed that whilst UV irradiation of the sea water appreciably reduced the bacterial load therein, it had no effect on either the load or the flora of the oysters.

Where spoilage patterns have been studied they have been typical of those described for other fish, being predominated by Gram-negative organisms of the genera *Pseudomonas, Moraxella* and *Acinetobacter*.[70,71]

As described earlier for fish, shellfish harvested from tropical waters have a longer shelf life than those from temperate waters. While both have a similar flora, the effects of ice storage are much more dramatic in the case of tropical fish. Shewan[23] attributed this effect to the fact that the resident flora was essentially mesophilic in the tropical fish and psychrophilic in those caught in temperate waters. This results in the former having a considerably longer shelf life in ice. Similar results were reported for shrimps by Cann,[67] Vanderzant *et al.*[77] and Christopher *et al.*[78]

A number of studies have examined the effects of shellfish processing on the spoilage flora and on organisms of public health significance, Cann,[67] Lee and Pfeifer[71] and Münzner[82] investigated the effects of washing, icing, cooking, peeling and breading in shrimp processing. Crabmeat production was studied by Lee and Pfeifer,[70] Ray *et al.*,[83] Ward *et al.*,[84] Phillips and Peeler[85] and Loaharann and Lopez.[86] In general the numbers of bacteria fall during washing, cooking and freezing and rise during handling, peeling, breading and packing. Recontamination after cooking was rapid and heavy in most processes. Lee and Pfeifer[70,71] showed that with shrimp and crab, *Moraxella, Pseudomonas, Acinetobacter* and *Flavobacterium* species originated from the new material and grew during refrigerated storage, *Arthrobacter* (coryneforms) and *Bacillus* species also originated from the raw material but did not grow during storage whereas *Micrococcus, Staphyloccus* and *Proteus* species were introduced during processing but again did not proliferate during storage. Cooking of both shrimp and crab kills most of the bacteria present, recontamination rapidly results in numbers approaching, or even exceeding, the initial contamination. Similar results were found by Ward and Tatro[87] and Ward *et al.*[84] in crab processing. The bacteriological status of such products will depend to a large extent on the hygienic state of the processing line, a number of the studies quoted demonstrated that significant improvements could be achieved by good hygienic processing.

Fish Mince

Mechanical fish flesh separating (deboning) machines are now increasingly used, not only to obtain a higher yield of flesh from traditionally exploited fish but also to utilise species of fish which were formerly considered to be unsuitable (uneconomic) for commercial processing. The comminuted product has spoilage characteristics which differ from those of chilled whole fish or fillets. Minced fish is the basis of traditional food dishes in some areas such as Eastern Europe, Scandinavia and Japan. Transformed into a form of fish paste—'kamoboko'—it is a long established favourite item of diet in Japan where, more recently, composite fish products such as fish sausage are produced in increasing quantity.[88,89]

Information on minced fish processing worldwide was collated by Bond for FAO[90] and the progress of the technology and product utilisation was assessed in the reports of an international conference in 1976.[91] Industrial experience in Canada was reviewed by Blackwood,[92] who considered that a microbiologically satisfactory product could be achieved by good quality control and efficient plant sanitation. He expressly noted that the deboning procedure should not be used as a salvage operation for unacceptable raw material. Crabb and Griffiths[93] reported on the bacteriological analyses of a wide range (1607 samples) of commercial frozen minced fish in relation to ICMSF standards[94] for these products. They concluded that, if the ICMSF standards were to be met, only good quality fish could be used and that high standards of hygiene are necessary throughout processing. Licciardello and Hill[95] found that 208 samples of blocks of frozen mince made from various fish species complied with these standards.

Cann and Taylor[96] examined the bacteriological quality of minces, from both white and fatty fish, prepared and chill-stored under experimental conditions. As had been the case with the frozen samples mentioned above, when good quality cod was used, fillets provided minces of the best quality. Trimmings, frame and back bone, in that order, produced minces which had progressively higher bacterial counts. When poor quality fish was used there was little difference in the bacterial counts on the resultant minces. These workers again found that minces would be required to be made hygienically from good quality fish if the final product were to meet ICMSFs standards. They also found no increase in total counts of bacteria when minces were prepared from good quality fillets. On the other hand, Raccach and Baker[97] claimed that mechanical deboning of either frames or headed and gutted fish increased the bacterial load by a factor of ten.

Minced fish, even that of good initial quality has a very short shelf life at chill temperatures and the usual form of storage is by freezing. In Japan minced fish is transformed into a stable product in the form of fish sausage.[88] This is a composite product which includes starch, lard, gelatin, flavourings and spices and owes its stability to the pasteurisation effect of the various heat treatments applied (80–90 °C for 40–60 mins), the action of permitted preservatives and the impermeable nature of the sausage casing. The usual form of spoilage in such products is due to the action of *Bacillus* spp.[98] *Bacillus* spp. have also been identified by Mori *et al.*[99] as the cause of a softening of spoilage of kamoboko, the contaminating vehicle being potato starch. Gram-negative rods of various types were considered to be the agents of a different type of spoilage—a brown discolouration—in kamoboko[100,101] and frozen surumi.[102]

Whilst the fish sausage is not established as a commercial product in the Western world, some experimental products have been investigated bacteriologically. King *et al.*[103] examined products made from skipjack tuna and striped marlin and found 300 organisms/g throughout storage in samples which were stable for 15–26 weeks at chill temperatures (35 °F, 45 °F). They pointed out that such products should be boiled before serving to obviate the botulism hazard.[104] Daley *et al.*[105] found that a sausage-like development product made from an under-utilised fish (mullet) had a short shelf life (two weeks at 2 °C) and off flavours appeared when bacterial counts at 35 °C reached 4.8×10^5 organisms/g. They concluded that some additional factors, such as impermeable casings, preservatives or vacuum packaging would be necessary in order to obtain a marketable product.

The results of studies on the mincing process show that bacterial numbers can increase rapidly unless strict hygienic procedures are observed. Subsequent spoilage of the mince at chill temperatures is more rapid than that of whole fish or fillets. This is presumably due to the intimate mixing of the flesh and bacteria normally resident on the outer surfaces coupled with the liberation of nutrients from ruptured tissue cells.

Miscellaneous Studies

Herrings and similar 'fatty' fish used for fish-meal and oil production are usually stored in bulk in tanks without refrigeration. Anaerobiosis presumably develops under such conditions and the bacteriology and biochemistry of spoilage under these conditions of storage has been investigated by Strøm and Larsen[106] and Strøm *et al.*[107] The growth rates at 15 °C of three organisms (*Enterobacter*, *Proteus* and *Aeromonas* spp.) which were isolated by an enrichment technique, and conversions of chemical components in herring extracts were studied. TMAO was found to be a limiting factor in the growth of the *Proteus* and *Aeromonas* cultures but not in the case of the *Enterobacter* strain.

An unusual aspect of the bacteriology of marine fish is the report by K. Schrøder *et al.*[108] of the isolation of psychrotrophic lactobacillus strains from the gut of saithe (*Gadus vireus*), capelin (*Mallotosis villosus*) and Norwegian fiord Krill (*Thysanoessa* sp.). The cultures were isolated by a special enrichment technique. The organisms, which resemble *L. plantarum* exhibit antagonistic activity against catalase-positive organisms. The isolation of lactobacilli from herring was previously reported by Kraus.[109] In the past there had been interest in the use of lactobacilli in the preservation of fish to be utilised for animal feedstuffs, etc. and the potential of this method of preservation has been re-examined by Lindgren and Clevström.[110]

PRESERVATION OF FISHERY PRODUCTS

Freezing and Cold Storage

The preservation of seafoods by freezing and cold-storage together with the distribution of frozen products by efficient cold-chains has become an accepted part of Western life. The 'consumer barrier' which, initially, was encountered by all frozen-food products has disappeared together with many of the myths about the quality of these products. This method of preservation and distribution is the most effective way of transporting seafoods to inland areas as it obviates losses due to bacterial spoilage. Most of the spoilage problems encountered in frozen fishery products are due to the quality of the raw material, to dehydration and fat oxidation during storage or to mishandling (e.g. accidental changes in temperature) during distribution. Freezing kills a proportion of the bacterial population in the fish; cold-storage and eventual thawing result in further, less dramatic reductions in numbers but it is, in practical terms, difficult to separate the effects of each of these factors. The significance of the total reduction in the numbers of bacteria so far as the spoilage of the thawed product is concerned is doubtful. Spoilage of thawed-out fish follows a similar pattern to that of chill-stored raw material. Once fish is thawed-out and conditions are favourable for bacterial growth, the time required for the population to reach the same level as that in the prefrozen material is relatively short.[5]

Experimental data on the interrelated factors, such as pH, salt concentration, water activity and cold shock, which govern the survival and growth of bacteria at low temperatures have been reviewed.[111,112] In a comprehensive monograph Calcott[113] summarised the results of the previous twenty years basic research on the effects of freezing and thawing on microbes and applied work carried out to solve the problems in industry. For the food microbiologist, the salient point is that, after freezing and thawing, some of the surviving cells are metabolically injured and are more fastidious than the original cells.[114] These stressed cells are capable of recovery but special methods are required to allow for this recovery to take place in order to detect all of them when investigating the bacteriology of frozen foods.[115,116,117] This and other aspects of the fate of bacteria in frozen foods has been dealt with in some detail by Splittstoesser,[118] Speck and Ray,[119] Kraft and Rey[120] and Michener.[121] The recovery of stress-injured cells is always important but it is particularly so when selective media are to be employed for the detection of them, otherwise the numbers of organisms will be underestimated. Mossel et al.[122] compared the effects of liquid media repair and the incorporation of catalase[123] in selective media for the recovery of freeze-stressed cells of Enterobacteriaceae both in pure culture and in frozen foods. They conclude that the use of two hours resuscitation in rich nutrient broth at room temperature (17–25 °C) is the method of choice, resulting in optimum recovery without leading to multiplication of the cells of the Gram-negative organisms under study.

Campello[124] reviewed the literature concerned with the bacteriological quality of frozen seafoods, including the influence of temperature and storage time, and concluded that the data from various countries were in close agreement. Gjerde,[125,126] who examined commercial frozen raw fish and Baer et al.,[127] who examined a wide range of commercial frozen, raw and frozen 'convenience' seafoods found that these products met with the bacteriological specifications set for them. Extensive bacteriological investigations of commercial fish frozen-at-sea and the effects of thawing on the bacterial load of fillets produced from them have been carried out in Canada.[128,129] The frozen-at-sea products had bacterial counts which would be expected in a high-quality product. Two thawing processes were studied and these had negligible effects on the total viable counts at 25 °C and little effect on the level of coliform organisms which were low in numbers throughout. In the authors' laboratory (unpublished data, 1973), it was found that the effect of thawing on bacterial counts of fillets and of gutted and headed fish was negligible in one type of thawing equipment whereas in a different machine, which was difficult to clean, contamination occurred resulting in increased counts. The effects of thawing, subsequent refrigerator storage time and culinary procedures on frozen seafoods in restaurant premises were investigated by Venkataramaiah and

Kempton.[130] They showed that if the thawing process is prolonged more than is necessary bacterial multiplication occurs rapidly. As a result of their studies they were able to formulate operational guidelines for this type of catering operation.

It is perhaps pertinent to note that current practices in freezing and thawing are designed to produce the best eating quality of the food and these conditions are generally the best for the preservation of the bacteria therein.

Salt Curing and Fermentation

Preservation of fish by salt curing and fermentation cannot be separated since most practical processes involve both to varying degrees. Throughout the world there is a wide range of products all of which are the result of the preservative action of salt and low pH together with varying degrees of breakdown of the fish by fermentative processes. An exhaustive review of these products was published by Mackie et al.[131] Microorganisms play varying roles in the different processes, desirable changes include achieving the right degree of degradation and producing the appropriate flavours, undesirable changes result in spoilage. The distinction between these is not always clear-cut, desirability being a subjective assessment. A desirable flavour in one part of the world is often considered a spoilage flavour in another and in any event spoilage changes are frequently a result of desirable changes proceeding too far.

The fermentation processes can be described firstly as hydrolysis of proteins to amino acids and peptides and conversion of amino acids to other compounds, many of which are highly flavoured. Final products are extremely varied ranging from whole fish to pastes and sauces. Prevention of spoilage is achieved by reduction of water activity (a_w) using salt, sugar and drying combined with a reduction of pH.

Traditional European salt cured fish products represent one end of the scale where there is little fermentation and microorganisms, predominantly bacteria, are thought to be involved only in the production of desirable flavours. The bacteriology of salt cured fish production[5] and the microbial safety of such production has been reviewed.[132]

There are essentially two types of product where the preservative action of salt is the predominant process and fermentation is thought to be involved only in flavour development. Dry salting is used for non-fatty fish and pickling for the fatty species. Variations in pickled products are achieved by the use of spices, vinegar and other additives. Descriptions of the processes can be found in the reviews quoted.[5,131] The microbial flora is derived from the fish itself and the salt or brine used. The normal Gram-negative spoilage flora of fish is not halotolerant and is replaced by halophilic and halotolerant micrococci, yeasts, spore formers, lactic acid bacteria and moulds. Little recent work has been done on this aspect of salt cured fish. Hamed et al.[133] gave data on lightly (12–17 % w/w) and heavily (15–20 % w/w) salted fish which confirmed earlier work and showed that the organisms which survived and multiplied during the process belonged to the genera *Micrococcus, Bacillus* and *Sarcina*. Fujii et al.[134] showed that in salted mackerel stored at low temperatures yeasts became the predominant organisms and the salt effectively inhibited trimethylamine-producing strains of *Vibrio* and *Aeromonas* genera. Fujii[135] also showed that a *Corynebacterium* species and a *Pseudomonas* species isolated from curing brines had an unusually high requirement for peptone (2·5–5·0 %) which is reminiscent of the increased nutritional requirements of stressed organisms (see freezing). Because the flora of these products is essentially halophilic, Gram-positive and mesophilic, few spoilage problems are encountered when the products are stored below 5 °C. At higher temperatures or if processing has been inadequate spoilage can occur rapidly. Insufficient salt in the flesh can result in spoilage by the usual Gram-negative bacteria. In other cases two types of spoilage organisms commonly occur. These are the extremely halophilic bacteria which cause a condition known as 'pink' and the halophilic moulds which cause a condition known as 'dun' or 'mite'. Both groups of organisms and the spoilage which they cause were fully described by Shewan and Hobbs.[5] The pink halophilic bacteria are actively proteolytic, have an optimum temperature of 35 °C to 40 °C but will grow between 5 °C and 50 °C and growth occurs over the pH range 6·0–10·0. They are obligate halophiles and require high concentrations of salt, growing actively in saturated brines. A variety of compounds such as hydrogen sulphide and indole, normally associated with spoilage, are produced as a result of their activities.

The taxonomic position of the pink halophilic bacteria is somewhat confused and they have been grouped together mainly because of their halophilic properties.

Data obtained in recent taxonomic studies[136] showed that only two genera appear to be substantiated: *Halococcus* which includes all the halophilic cocci in one species, *H. morhuae*, and *Halobacterium* which comprises all the rod shaped organisms. This latter genus can be differentiated into two species, *H. salinarium* and *H. cutirubrum* by certain biochemical tests as well as salt/temperature growth ranges and lethal salt/temperature ranges.

The proteases of *H. salinarum* were studied by Koveleva et al.[137] who found that maximum activity occurred at pH 7·0 and 5 M NaCl with one strain, and pH 6·5 and 4 M NaCl with another. Lipase activity in strains of *H. cutirubrum* and *H. morhuae* has also been described.[138]

A quick salting technique of producing a heavily salted minced fish has been described.[139-142] The finished product contains 24–26% salt, 40–45% moisture and 30–35% fish solids giving a water activity of 0·745. Varga *et al.*[143] investigated the microbiology of this product and concluded that spoilage by halophilic bacteria could be controlled by reducing the water activity to 0·70 or by the addition of 0·3% sorbic acid.

The halophilic moulds, which belong to the *Sporendonema* and *Oospora* groups, have not been studied as intensively as the halophilic bacteria. They require at least 5 to 10% salt but grow actively in concentrations up to 20%, have an optimum temperature of 30 °C and fail to grow below 5 °C. Growth occurs over pH ranges of 3·3 to 7·4. They do not decompose salt fish or produce off odours but their presence detracts from the value of a product merely because of their undesirable appearance.

Being strictly halophilic, the pink bacteria are readily killed by exposure to freshwater and control is best achieved therefore by good hygiene and refrigeration. Since they originate in solar salts the choice of a solar salt with a low incidence of those organisms or the use of a mined or vacuum dried salt is an effective preventative measure. The spores of halophilic moulds are not affected by exposure to water but as with 'pink' bacteria one method of control is refrigeration. Doesburg *et al.*[144] showed that both could be controlled with sorbic acid, dipping the fish before salting was most effective.

Food poisoning seldom occurs with this type of product but may be caused by *Staphylococcus aureus* which can grow in some of them. Bacterial production of dimethylnitrosamine has been demonstrated in salt fish and nitrate reducing bacteria, including *S. aureus*, have been incriminated.[145,146]

The various highly salted and dried products of Scandinavia and Europe in which some decomposition has occurred represent the next stage to salt curing in that some degree of fermentation is allowed to take place. The process is controlled by combinations of salt, sugar and acid. In some cases preservatives such as sodium nitrate and benzoic acid may be used. Such products include the French anchovy, Scandinavian Tidbits and anchovies and Surstrøming and Rakørret products. In the case of these last two, considerable putrefaction has occurred. Hansen[147] has described a new method for bottling this type of product that permits a storage life of up to two years.

In South-East Asia fish pastes and sauces are of greater importance. In this type of food, fermentation proceeds much further than in the products already described. Both proteolytic enzymes, either added or already present in the fish, and microorganisms are involved in the fermentation which again takes place in the presence of high salt concentrations. Many of these products are made without starter cultures or added carbohydrate, the process relying on salt, the flora which develops naturally and possibly added enzymes to achieve the desired decomposition and preservative action. Lee,[148] describing the microbial flora of Korean seafood pickles, found a wide range of organisms including *Micrococcus, Brevibacterium, Sarcina, Leuconostoc, Bacillus, Pseudomonas* and *Flavobacterium* spp. In addition he described enzymes, of both fish and bacterial origin, active in the process. These included proteases, RNA depolymerase and diesterases. Orillo and Pederson[149] investigated a fish-rice (burong dalag) from the Phillipines. As might be expected, because of the additional carbohydrate, a lactic acid fermentation occurred and the pH rapidly fell to below 4·0 after one week. The main organisms responsible were *Leuconostoc mesenteroides, Pediococcus cerevisiae* and *Lactobacillus plantarum*.

Sands and Crisan,[150] in a microbiological study of fermented Korean seafoods, confirmed some earlier findings. A wide range of bacteria including *Bacillus, Micrococcus, Pediococcus, Pseudomonas, Serratia* and an occasional *Clostridium* spp. were present as well as a few yeasts but no filamentous fungi or *Actinomycetes* were detected.

Salt cured, smoked and fermented fishery products are traditional foods which tend to be associated with regional or ethnic groups. In the UK and other developed countries salting and smoking are nowadays used not so much for their preservative action but mainly to cater to particular tastes. Other factors have had an influence on the production of some varieties; increased costs of fishing mean that salt-cod or salt-ling cannot always be produced as a cheap item of food as it was formerly. Dwindling stocks of herring in UK waters mean that various curing or pickling processes using this fish have almost totally disappeared. However, in developing countries salt curing and smoking processes still have potential especially in the exploitation of new fisheries. Fermentation is used extensively in different parts of the world not only as a means of processing fish especially in hot climates but also for economic and dietary reasons. Some of these processes might be used as a means of exploiting new fishery resources.

Smoke Curing
Smoke curing is a traditional preservation process which developed alongside salt curing and fermentation and in fact relies to some extent

upon salting. Four basic treatments are involved—brining, drying, heating and smoking—all of which influence the bacterial population and its subsequent development during storage. The ability of food poisoning organisms to grow in the final product will be considered elsewhere. Variations in these four treatments and the use of a wide range of fish species have resulted in a very large variety of smoked products. Although smoke curing was developed primarily as a preservative process, with the wider use of chilling and freezing the cosmetic and organoleptic properties of the products are probably more important now. There are wide variations in amount of smoke, heat and drying during processing and in the salt content of the final product. In lightly cured fish such as kippered herring or finnan haddock the salt content is 1·5 to 2·5%, smoking is light (about four hours in a Torry Kiln), the temperature of the fish rises to about 30 °C and a weight loss of approximately 20% is achieved. At the other end of scale are the 'red' herring with a salt content of 10 to 12% and the hot smoked products which are heated to a temperature of 60 °C during smoking and are virtually sterile immediately after the smoking process.

Since the review by Shewan and Hobbs[5] little has been published on the spoilage flora of smoked fish. Gram-negative organisms are more sensitive to the effects of salt, smoke, heat and drying than are Gram-positives. In lightly cured products the process has a relatively small effect and a typical Gram-negative spoilage flora develops on storage. In the case of the more heavily cured products, the Gram-negative organisms are killed leaving an essentially Gram-positive flora, the heaviest cures resulting in an almost sterile product as already mentioned. The better storage properties of these more heavily cured products are due to the greater overall reduction in the numbers of bacteria and the fact that the residual Gram-positive organisms do not grow as quickly as the Gram-negative organisms at refrigeration temperatures. The work of Lee and Pfeifer[151] on smoked salmon bears out this conclusion: they found that the aerobic flora at the retail stage was essentially Gram-positive. This product was cold smoked, had a salt content of 3·2 to 8·2% and a moisture content of 48 to 64%. The results of Deng et al.[152] on smoked mackerel also showed these effects.

The effects of packaging smoked salmon have been examined chiefly from the point of view of hygiene, food poisoning or extension of shelf life (vide infra). However, one investigation has shown that in packaged smoked fish Gram-positive cocci predominate the spoilage flora.[153]

It has long been recognised that the preservative action of smoke is due to the phenolic fraction.[5] Recent work on the development of 'liquid smoke' has concentrated on retaining the preservative action and removing the potentially carcinogenic fractions.[154,155] Using these liquids the fish are dipped in smoke solutions rather than having smoke deposited in the usual way. Smoked fish products made in this way have not yet attained widespread commercial application and would require some degree of heating and drying in order to simulate traditional smoked fish.

Irradiation

It has been established for some time that irradiation sterilisation of fish and fishery products is not a feasible commercial proposition using the techniques presently available. This is mainly because the irradiation doses required (4·0 to 6·0 M rads) produce undesirable flavour changes. On the other hand irradiation pasteurisation, using doses in the region of 0·3 M rads or less, can give a two or three fold extension of the shelf life of many fishery products without perceptible alteration in the organoleptic properties.[156] In spite of these findings the process is not used commercially on any large scale at present. The main reason for this is the necessity in most countries for National legislation to permit the process as safe. The main requirement is to establish that the process does not introduce an increased health hazard. Most of the problems in this area are concerned with the possibility of creating mutagenic or carcinogenic components in the food as a result of irradiation. Whilst much progress is being made in this direction it is outside the scope of this review and only the microbiological aspects of irradiation will be discussed.

A number of international conferences were devoted specifically to the question of food irradiation between ten and fifteen years ago[157,158,159] and in addition it has been the subject of joint IAEA–FAO expert panels.[160–163]

Bacteriologically the effect of sub-sterilisation doses of irradiation on fish is to kill a large proportion of those bacteria which normally cause spoilage, thus achieving an extension of the shelf life. As described elsewhere in this review, bacteria of the genera *Pseudomonas* and *Alteromonas* are chiefly responsible for the spoilage changes of iced fish. It is precisely these organisms which are the most sensitive of those present to irradiation. The predominant organisms surviving irradiation pasteurisation are *Moraxella, Acinetobacter, Micrococcus* sp. and members of the coryneform groups. These last two do not grow well at 0 °C, hence the main spoilage flora of irradiation pasteurised, iced fish is composed of *Moraxella* and *Acinetobacter* species. Much of this early work carried out on relatively few species of fish (cod, haddock and herring) was summarised by Hobbs and Shewan.[156] Other publications since then have generally confirmed this work.[164-177] These publications have included a variety of marine fish, shellfish and freshwater fish. Minor variations are reported in the details of the surviving flora, and varying doses from 0·05 to 0·5 M rads have been used. The importance of this work is that the main spoilage bacteria, *Pseudomonas* and *Alteromonas* spp. are greatly reduced or eliminated. Surviving Gram-negative flora (*Moraxella* and *Acinetobacter* spp.), whilst able to grow well at ice temperature produce much less offensive organoleptic changes and the Gram-positive flora grows poorly, if at all, at ice temperature. If irradiation is combined with vacuum packaging the *Moraxella* and *Acinetobacter* species are further inhibited by the lack of oxygen and a predominantly Gram-positive spoilage flora (coryneform, *Lactobacillus*, *Micrococcus*, *Sarcina* and occasionally *Bacillus* spp.) develops.[164,169,178] Thus, because of the variability of types of organisms which survive in radiation pasteurised fish, a number of authors have reported that total viable counts are of little value in assessing the quality of end products.[170,172,179] On the other hand, Miyauchi[170] claims that at doses in the region of 0·05–0·10 M rads some irradiated fish are spoiled when the bacterial count reaches 10^8/g as opposed to 10^6/g for unirradiated fish.

One of the possible health hazards which could increase as a result of the irradiation pasteurisation process is that of botulism. As described elsewhere in this review, non-proteolytic strains of *Cl. botulinum* can be present in freshly caught fish. Irradiation at up to 0·5 M rads has little or no effect on the spores of this organism but extends the shelf life of the fish. A considerable amount of work has been done on this subject and it was exhaustively reviewed by Hobbs.[180] It is clear that under certain circumstances, namely a relatively high contamination level and storage above refrigeration temperatures, irradiated fish can become toxic before the fish is organoleptically objectionable. This situation can arise where poor hygiene and process control exist; it must be pointed out, however, that in this respect irradiation is no different to some smoke curing processes and heat pasteurisation.

Irradiation could, of course, be used in conjunction with other preservation processes and there is some evidence of synergism. Combinations of γ-irradiation and UV radiation or chemicals such as sodium chloride, nitrite and nitrate are reported to give better preservation than would be expected from either treatment alone.[181-184]

Packaging

Packaging of fish offers several advantages; it protects the fish from contamination and if an oxygen-free atmosphere or vacuum pack is used it inhibits fat oxidation. It can, in some cases, provide an extension of shelf life and has other commercial advantages in that the product can be made more attractive and be handled more easily.[185] Bacteriological considerations only arise with unfrozen fish and where oxygen permeable packs are used these are no different from those with unpackaged fish. In cases where low permeability materials are used either with the application of a vacuum or a modified atmosphere such as carbon dioxide, then special bacteriological considerations exist. Firstly, because of the limited supply of oxygen or its complete absence, many of the normal spoilage bacteria (*Pseudomonas* and *Alteromonas* spp.) are inhibited. Generally, those bacteria which do grow well because of the anaerobic conditions produce different metabolic results. Vacuum packaging does in fact result in an extension of shelf life though this is usually not large.[186–191]

Because of the anaerobic conditions in vacuum packs the possibility of encouraging the growth of *Cl. botulinum* must be considered. The early evidence showed that growth and toxin production in vacuum packed fish was only slightly faster than in unpackaged fish.[192,193] Further work has been carried out on a wider range of fish and this has confirmed the earlier findings.[194] There has been a growing interest in the use of varying levels of carbon dioxide in packaged food.[195,196] These studies indicate that an extension of shelf life can be achieved and that this is due to inhibition of the normal spoilage bacteria, firstly by the lowered oxygen level and, secondly, by the toxic effects of carbon dioxide itself.[197]

Again the question of growth of *Cl. botulinum* arises, and it has been reported that, as with vacuum packs, the encouragement of growth and toxin production is marginal and of no commercial significance.[198]

HYGIENE AND FOOD POISONING

Quality Control and Specifications

The overall objective of food hygiene and quality control is to present the food to the consumer in a wholesome state and without any health hazard. Historically, the prime concern was with health hazards; however, today prevention of spoilage is also a major concern. Fish and fishery products are amongst the least stable of foods and spoilage occurs relatively rapidly. Because of this there has been widespread use of traditional preservation methods such as salting, smoking, chilling and fermentation. The use of these methods of preservation, coupled with the fact that fish caught in the open sea are usually free from food poisoning organisms, are the reasons that this food material generally has a good reputation as far as human health hazards are concerned.[5,199] However, two types of food poisoning bacteria are present in the marine environment and have caused serious problems. These two organisms, *Clostridium botulinum* and *Vibrio parahaemolyticus* will therefore be discussed in more detail later.

It has become increasingly obvious, however, that in some parts of the world commercial fisheries are close enough inshore to be affected to a significant extent by contamination from sewage outfalls.[200–204] Freshwater fish can be contaminated with Salmonella.[205] Contamination of fish with organisms of public health significance nevertheless remains primarily a problem of handling and processing. A number of studies in recent years have borne this out and moreover have shown that good hygiene procedures result in beneficial effects.[87,128,206–210]

With a few minor differences, the control of spoilage bacteria requires much the same procedures as the control of those presenting health hazards. Hygienic procedures therefore are designed to prevent contamination of the fish with undesirable organisms and to prevent, as far as possible, the growth of those organisms which are already present. Quality control procedures and microbiological standards or specifications are best viewed as measures to ensure that the hygienic procedures used are effective. Bacteriological tests can seldom be used for acceptance or rejection because of the time required to obtain the results; an exception to this is where frozen foods are concerned. They are, however, useful as a test of the effectiveness of hygienic procedures and in devising adequate cleaning and processing.

There has been considerable interest in devising more rapid tests to overcome the time problem;[211,212,213] however, at present, traditional tests and methods are still in general use. In using traditional tests many factors, including the composition of the food being examined, can influence the results obtained. There have been efforts to examine the effectiveness of bacteriological tests with fishery products and to make the tests less time consuming. Anderson and Baird-Parker[214] included fish when they evaluated a rapid direct plate count method for *E. coli* type I. Using indole production and tryptone bile agar counts were successfully obtained in 24 h. Andrews *et al.*[215] also described a rapid method for recovery of *E. coli* but this time from sea water from which shellfish were harvested. They described two media which gave results in 24 h both of which were a slight improvement in efficiency on the standard American Public Health Association method which takes 72 h. Francis and Twedt[216] adapted a pour plate method and using a combination of pH 8·0 and incubation at 41–44 °C were able to count faecal coliforms in shellfish after

72 h. A further variation proposed for detecting faecal coliforms by incubation of MacConkey broth cultures at 37 °C for 2 h followed by incubation at 44 °C for 24 h, gave results equally as good as the standard MPN methods with oysters.[217] For *E. coli* type I a peptone water medium was also incubated for subsequent indole testing. In the examination of minced fish Zaleski and Fik[218] recommended a pre-incubation in a trypticase–soy–yeast extract broth followed by plating on selective media to obtain results within 24 h for both *E. coli* and *Salmonella*. Reduction of the test time to 24 h for total plate counts, coliforms, *E. coli* and *Staph. aureus* was achieved by using modifications of normal media and methods.[219] In examining fish and shrimp for *Staph. aureus* a slide coagulase method has been described which permitted detection within 24 h.[220]

A new medium was proposed for the rapid detection of *Cl. perfringens*[221] and was used successfully by Kanzaki *et al.*[222] in a MPN method for the examination of raw and processed fish products. While it is obviously desirable to have rapid methods there is less information available regarding the value of total counts and various indicator organisms for predicting health hazards or excessive contamination in fishery products. Chang and Choe[223] found that faecal coliform counts were more useful indicators of hygienic quality than enterococci when examining raw fish at the retail level and that there was no correlation between total counts and counts of any indicator organisms. Andrews *et al.*[224] found that as an indicator for the presence of *Salmonella* both total coliform and faecal coliform estimations were satisfactory though high numbers of either did not necessarily indicate high numbers of *Salmonella*. Ayres[225] examined combined samples of oysters, mussels and clams and found good correlation of *E. coli* counts with *Cl. perfringens* counts but not with numbers of coliform and faecal streptococci. Again total counts did not correlate with counts of indicator organisms but he suggested that the relationship between total counts at 20 °C and 37 °C might be useful with raw products. This was suggested earlier by Hobbs[199] who quoted an example in the processing of scampi (*Nephrops norvegicus*) where changes in the relationships of these counts could indicate where contamination was occurring. It is clear from the information available that total counts at 37 °C by themselves offer no useful information regarding contamination levels or health hazards. This is not necessarily the case on a particular production line where the normal level in the product is known. In this situation an abnormally high count does indicate that something has gone wrong. Likewise where the count at 20 °C is regularly monitored it will rise with a given product during chill-storage in a predictable way.[199] Indeed, with chill-stored fishery products the only bacteriological test which gives any information regarding the spoilage status or potential shelf life is the total count at 20 °C. All the other bacteriological tests relate to potential health hazards. In the case of these tests the results indicate that there is a specific, preferably low, probability of a health hazard and the aim is to get this probability as low as possible. Amongst the factors affecting this is the sampling rate. Bacteriological tests are generally destructive and this combined with restrictions imposed by available manpower and cost are the chief limiting factors in any quality control programme.

Many national and international bodies have been engaged in work related to standards, specifications or guidelines though there remain divergent opinions regarding the practicality or value of them.[226–238] At international level the Codex Alimentarius Commission under the auspices of WHO and FAO have attempted to establish guidelines for the development and application of microbiological specifications for foods. Three useful categories of microbiological criteria were defined by the second Joint FAO/WHO Consultation on Microbiological Specifications for Foods. These were advisory guidelines, specifications attached to codes of hygiene practice and mandatory standards. Essentially, 'standards' have legal backing, 'guidelines' are similar but not enforceable by law and 'specifications' are commercial agreements. In addition the International Association of Microbiological Societies' (IAMS) International Commission on Microbiological Specifications for Foods (ICMSF) tackled the particularly difficult problems of formulating standards. Sampling plans and specifications were proposed for a wide range of foods.[94,239] The main difficulties in using microbiological standards as part of a statutory end-product specification are that sampling intensity and costs will be very high if results are to be relied upon and the results of the tests available cannot always be related to health risks or eating quality.[235,238] The general consensus of opinion is that to ensure the safety of foods and maintain good eating quality, control of hygiene is necessary at all stages of production and distribution and standards should be employed only in cases where a high risk has been identified and there is a real chance of reducing a health risk or improving eating quality. In exercising control of hygiene it will, of course, be useful to employ microbiological criteria to ensure that procedures are effective.

Shewan[228] presented a detailed discussion of such microbiological standards as existed for fish and fishery products. Many guidelines and specifications are in commercial use but those quoted in the ICMSF publication,[94] however, are typical. An example of a well identified risk was quoted by Shewan[236] where cooked prawns imported into the UK had caused food poisoning. A microbiological specification was agreed between the importers and the Public Health Authorities which was designed to alleviate this specific problem.

A detailed examination of shellfish in international trade carried out by Cann[67] demonstrated how bacteriological specifications could be used during processing and distribution to ensure safe and good quality products in international trade. Several examples of such specifications were quoted.

An older example of the use of bacteriological specifications is the control of molluscan shellfish where previously typhoid had repeatedly occurred as a result of their consumption. In this case the shellfish were harvested from polluted waters and the standards applied related to the water rather than to the shellfish themselves. Ayres[225] has presented details of the existing situation with bivalve molluscs on the UK market.

Unlike warm-blooded animals, fish do not have a specially adapted intestinal flora. The bacteria present in the intestines and on the surfaces of fish reflect the flora of the feed and environment. This presents special problems when considering specifications for international trade.

As with other foods the purpose of any particular standard or guideline for fishery products must be clearly understood. As an indication of any possible health hazard the specification will usually include total numbers of aerobic bacteria growing at 35–37 °C, numbers of indicator organisms such as coliforms or *E. coli* and counts of specific pathogenic bacteria such as *Salmonella* and *Staphylococcus aureus*. Such a specification will give no indication of the state of spoilage or potential shelf life of the product. At present no bacteriological test will give this information for fish or fishery products when used as an end-product specification. Regular examination of a production line can, however, give useful information though it will probably be cheaper and more effective to ensure proper control of processing parameters, particularly temperature.

Vibrio parahaemolyticus

This organism is unusual in that few marine bacteria are known to be pathogenic to man. It was first identified as the aetiological agent of a specific form of gastro-enteritis in Osaka, Japan in an outbreak of food poisoning in 1950 involving more than 270 cases.[240] In the next decade or so many outbreaks of food poisoning associated with the ingestion of seafoods occurred in that country where the causative organisms were identified as halophilic pathogenic organisms (*vide infra*). The original identification of this organism as the cause of a specific form of gastro-enteritis in man, and the history of the careful and intensive studies which established that it was the major cause of food poisoning incidents in Japan during the warmer months has been thoroughly documented.[241]

At first, identified outbreaks were confined to Japan but, by the late 1960s and early 1970s, outbreaks in other areas showed that this organism was of more direct importance to the Western world than had first been realised.[242,243,244] This was coincidental with an increase in international trade in fishery products, particularly in crustacean shellfish.

Since the middle 1960s it has become the subject of investigation by marine fishery, food and public health laboratories and in consequence there is a plethora of literature reporting studies on all aspects of this organism. In the original incident in Osaka the outbreak was traced to a semi-dried fishery product called 'shirashu'. Isolates from patients and from the food could not be identified as any previously described human pathogen and were named *Pasteurella parahaemolytica*. Subsequently, Takikawa[245] isolated a Gram-negative rod from an outbreak of food poisoning involving more than 120 cases in a hospital. The organism was isolated on a 4% NaCl nutrient medium (used to isolate Staphylococci) and was initially considered to be a member of the genus *Pseudomonas*; it was named *Ps. enteritis*. The halophilic nature of these isolates was realised (this had not been so with the *Pasteurella parahaemolytica* strains), and although the food vehicle implicated in the outbreak was brined cucumber, it was thought that the original source was mackerel which had contaminated the cucumbers. In the late 1950s several Japanese bacteriologists investigated strains of 'halophilic pathogenic bacteria' from sources such as sea water, bottom deposits, plankton, marine animals as well as from clinical sources.

In 1961, Miyamoto *et al.*[246] reported the results of comparative studies of the strains isolated by Fujino[240] and by Takikawa[245] as well as strains which they themselves had isolated from sea water and clinical sources. Based upon the halophilic character of these organisms and their fermentative action on glucose they proposed that they should be placed in a new genus *Oceanomonas*. From putrefying fish they had also isolated strains of a non-pathogenic halophilic organism for which they proposed the name *Oceanomonas alginolytica*. In 1963, as a result of taxonomic studies of over 1000 strains of 'pathogenic halophilic' organisms. Sakazaki *et al.*[247] proposed that because of their relationship to vibrios these organisms be called *Vibrio parahaemolyticus*. They recognised two groups,

TABLE 1

BIOTYPES OF *Vibrio parahaemolyticus*

Differential characters	Biotypes:	
	1	2
Growth in peptone water + 10% NaCl	–	+
Voges–Proskamer reaction	–	+
Sucrose fermentation	–	+

the enteropathogenic strains—biotype 1 and those of doubtful or no pathogenicity—biotype 2; the features separating the two biotypes are shown in Table 1.

Zen-Yoji *et al.*,[248] having carried out epidemiological and aetiological studies and numerical taxonomy of the organisms confirmed this division of grouping and showed that the incidence of biotype 2 in clinical material was low whilst there was a high incidence in fish, shellfish and catering equipment. The taxonomic studies of Fujino *et al.*[249] agreed with those of Sakazaki and his colleagues in showing a relationship of these organisms to the genus *Vibrio*. Subsequently, in 1968, Sakazaki[250] reported that his biotype 2 strains were identical to the cultures of *Oceanomonas alginolytica* described by Miyamoto *et al.*[246] and he suggested that biotype 2 strains should be recognised as a new species *Vibrio alginolyticus*. In an extensive study of organisms assigned to the genus *Vibrio* Colwell[251] also concluded that *V. parahaemolyticus* and *V. alginolyticus* were significantly different species.

Baumann *et al.*[252] included enteropathogenic strains of *V. para-haemolyticus* in a comparative study of marine bacteria isolated off the coast of Hawaii. In associated electron microscope studies[253] it was shown that whilst the cells of *V. parahaemolyticus* and *V. alginolyticus* possessed a single, sheathed, polar flagellum when grown in liquid media, on solid media the same strains possessed numerous simple peritrichous flagella as well as the polar flagellum. They proposed that these organisms be assigned to the redefined genus *Beneckea* on the grounds that there were significant morphological differences from the accepted definition of the genus *Vibrio* as well as some biochemical features including their chitinase activity. Baumann and Baumann[254] reviewed the role of these and related microorganisms in the marine environment but subsequently withdrew their taxonomic proposals[255] and agreed that the organisms should be considered as *Vibrio* spp. as had been proposed by the international sub-committee on the taxonomy of vibrios.[256]

V. parahaemolyticus is a motile Gram-negative rod which grows between 10 °C and 44 °C with optimum growth at 35–37 °C.[257] Growth occurs over the range of pH 5·0–11·0[258] with optimum growth at approximately pH 7·5. It has a requirement for sodium chloride, growing in concentrations of between 0·5 % and 8 % NaCl, the optimum being between 2 and 3 % and it fails to grow in the absence of NaCl.[258,259,260]

The generation time under optimum conditions is very short, generally in the region of 11 to 13 mins, though some strains have been reported to have generation times of 8 to 9 mins[261] and 5 to 7 mins.[262] The organism does not grow at temperatures below 8 °C to 10 °C and in fact dies off at chill temperatures (0 °C to 8 °C). It does, however, survive well in frozen foods.[67,263]

Cultures are biochemically active in a variety of laboratory test media unlike most of the psychrophilic commensal flora found on marine fish. In this respect it resembles *Vibrio angillarium* and the luminous vibrios but biochemically and morphologically it most closely resembles the luminous organisms which belong to the genus *Lucibacterium*.[253,264]

The importance of an awareness of this organism in modern fish technology is apparent from some of the features described. In warm coastal waters (20–25 °C) it is likely to be present in moderate numbers, it can proliferate rapidly on seafoods (and other foods) at ambient temperatures especially above 20 °C and, if ingested, has the ability to grow extremely rapidly in the human intestine. In man it causes a form of gastro-enteritis, accompanied by severe abdominal pain, the onset being sudden and the incubation period varying from 2 or 3 h to 24 or even 48 h.

It is appropriate at this stage to emphasise that it was some eighteen years before there was some clarification of the enormous problems which faced the original workers in isolating and characterising this organism. Selective and enrichment media were developed for the 'new' organism partly through work in this particular field, but also due to equally intensive efforts applied during the years 1963 to 1973 to investigations into the increasing problem of cholera-like infections. To the practical microbiologist the isolation, identification and enumeration of *V. parahaemolyticus* is still a complex and difficult procedure. Whilst appropriate media are now available it is certain that, in samples under test, *V. alginolyticus* will be present often in larger numbers than *V. parahaemolyticus*. The isolating procedures will enrich and selectively grow both these species and, it seems, an ever increasing range of other vibrios as well.[265] Furthermore, although it is possible to distinguish between most strains of *V. parahaemolyticus* and *V. alginolyticus* by the use of a limited series of tests (Table 2) during large

TABLE 2

CHARACTERISTICS IMPORTANT IN THE IDENTIFICATION OF *Vibrio parahaemolyticus* AND THE DIFFERENTIATION OF *V. parahaemolyticus* AND *Vibrio alginolyticus*

	V. parahaemolyticus	V. alginolyticus
Colonies on sea water agar (20°C and 37°C)	Discrete	Swarming occurs (some strains)
Colonies on TCBS[b] agar (37°C)	Green	Yellow
Sucrose peptone water (+2% NaCl)	No change[a]	Acid, no gas
Voges–Proskauer test	−[a]	+
Triple Sugar Iron agar—slant	Alkaline (red)	Acid (yellow)
butt	Acid	Acid
H₂S	−	−
Growth in 1% trypticase broth:		
without added NaCl	−	−
with 8% NaCl	+	+
with 10% NaCl	−	+

Both species share these features: Gram-negative, motile rods, produce pleomorphic forms on agar, sensitive to pteridine compound 0/129 (vibriostatic agent),[9] fermentative in glucose Marine O–F medium, no gas from glucose, indol positive,[a] urease negative,[a] growth at 42°C in 1% trypticase broth with 2% NaCl.
[a] Variations have been observed in some strains.
[b] Thiosulphate–citrate–bile salts–sucrose agar.

scale investigations a number of aberrant strains of both species will be encountered. Variations in urease activity, sucrose fermentation and indole production are not uncommon.

The use of special media for the isolation and detection of halophilic vibrios has shown that this group of organisms is more heterogenous and, in coastal zones, occurs in greater numbers than had previously been realised.[265,266] *V. parahaemolyticus* is essentially a coastal and estuarine organism and has only rarely been isolated from fish caught in the open sea.[263,267] Its growth depends not only upon temperature but is influenced by the presence of organic nutrients such as may derive from seafood processing or sewage.[263] Thus, although its incidence in water in coastal zones is usually seasonal, in estuarine waters rich in organic nutrient it is possible to detect it throughout the year.[266]

As already stated many investigations have been carried out into the incidence of *V. parahaemolyticus* and there are now many publications on this organism. It has been detected in all coastal zones investigated and its incidence is related mainly to the environmental temperature.[242,243,244]

Amongst the earliest comprehensive publications is the report of an International Symposium held in Tokyo in 1973.[241] This report is of particular interest for, as well as covering many aspects of the initial investigations in Japan, it reflects the commencement of the worldwide interest during the late 1960s and early 1970s. This interest was stimulated by reports of outbreaks, outside Japan, of food poisoning due to *V. parahaemolyticus*. As already mentioned, public health laboratories were at that time increasingly aware of the spread of cholera-like infections and methods for isolating vibrios were routinely employed in investigations in cases of gastro-enteritis. During this period there were investigations into the incidence of *V. parahaemolyticus* and reports of outbreaks of food poisoning due to this organism in many areas.[266-274] The interest continued throughout the 1970s with further publications[275-289] and several papers presented schemes for the isolation, identification and enumeration of this organism.[266,274,277,281-284,286,288,289]

Serological studies aimed at elucidating the epidemiology of *V. parahaemolyticus* food poisoning were carried out by Sakazaki *et al.*[290] and by the use of 24 'O' and 46 'K' antigens 54 different serotypes are at present recognised. The scheme is based on the serology of isolates from clinical material, many isolates from the environment and seafoods do not match any of these types. Clearly further work is necessary to clarify the relationships between these strains.

Whatever the serotype, however, most clinical isolates produce β-haemolysis in a medium developed by Wagatsuma[291,292] whereas cultures obtained from the environment or seafoods usually do not.[266,293] This haemolysin test is known as the Kanagawa reaction (Kanagawa is the name of the region in Japan where the test was developed). Barrow and Miller[266] stress the necessity of using this test under precisely defined conditions. They also refer to their previous studies in which Kanagawa-positive strains outgrew Kanagawa-negative ones in prawn-broth cultures under certain conditions and suggest that if this occurs in the human intestine it may be one explanation for the positive Kanagawa reaction of clinical isolates.

Beuchat[263] reviewed studies on environmental and food processing factors which influence the survival and growth of *V. parahaemolyticus*. Although it is extremely sensitive to heat and to chill temperatures wide ranges of tolerances have been reported. In the environment Kaneko and Colwell[294,295] showed that the organism survived only on plankton and in sediments during the colder months but as the temperature increased it was also detectable in the water. Similar observations have been made in ecological studies of the closely related luminous bacterium *Lucibacterium* (*Beneckea*) *harveyi*.[295]

Although *V. parahaemolyticus* is sensitive to chill temperatures (0–8 °C), it survives the freezing process and has been isolated from frozen fishery products.[67,263] Mishandling of batches of thawed-out products may lead to the growth of this organism in them or to cross-contamination of other products. This is particularly important in shellfish processing when heat treatments are used.[67] Because of the reduction in the numbers of the competitive flora as a result of heat treatment, substrate conditions are favourable for the very rapid proliferation of *V. parahaemolyticus* should it gain access and if temperature control is not increased. Some of the reported outbreaks of *V. parahaemolyticus* food poisoning have, in fact, been due to contamination of heat-treated shellfish which had been subsequently exposed to ambient temperatures.[297,298]

The survival of *V. parahaemolyticus* under conditions of stress has led to a number of investigations into the suitability of diluents, and both enrichment and selective media for the recovery and enumeration of organisms stressed by various treatments.[299–304] These factors should be taken into account when investigating any particular situation.

To the food processor the implications of the presence of *V. parahaemolyticus* in the natural environment varies not only with climate but with dietary habits.[305] In Japan, due to the climate and dietary habits coupled with the fact that seafoods form a high percentage of the protein intake, it is recognised to be a serious problem in the summer.[277] In Great Britain[266,273,275,279] and other European countries[276] it is not a serious problem but in some temperate zones, e.g., USA, outbreaks involving a large number of cases have occurred.[243,277]

Like many food poisoning hazards, it will always be a potential risk in temperate zones, during the summer months with locally caught seafoods and throughout the year with seafoods imported from warmer climates. This is particularly true with shellfish: the association of the organism with crustacea has been clearly established and filter feeding molluscs, of course, 'enrich' it from the surrounding water. Shellfish, transported in bulk in international trade, are thawed-out and refrozen in smaller units for distribution. There is no doubt that this form of food processing is a source of contamination, for instance, in Great Britain[266,275,279] (unacceptable levels of *Staphylococcus aureus* and the presence of enteric pathogens are, however, more important hazards in such products). Contamination may equally affect foods other than fishery products. *V. parahaemolyticus* has been isolated from a number of different foods[245,263] and has the ability to grow in a wide range of protein foods of relatively low salt content.[306]

Whether or not a search for *V. parahaemolyticus* is included in quality control procedures or in surveillance by health authorities will depend upon considerations such as those just discussed. Such a decision will depend upon a knowledge of the likely incidence of the organism in the raw material being processed and the likelihood of food poisoning arising from consumption of the product.

A final consideration is that organisms of the *V. parahaemolyticus–alginolyticus* group and similar, unidentified, halophilic vibrios have been involved in pathological conditions other than food poisoning. In mariculture they have been isolated from wounded and diseased species of fish and shellfish but there is little evidence that they are other than secondary invaders.[307] However, in humans, they have been isolated from ear infections in swimmers and from abrasion of the skin in swimmers, divers and people wading (e.g. in search of shellfish) and have also been of aetiological significance in a number of very serious clinical conditions.[308–313] These forms of infection should be borne in mind in relation to the health of personnel employed in fish handling and processing.

Clostridium botulinum

The association of certain types of botulism with fish was recognised before the causative organism was isolated and characterised. However, a sudden increase in outbreaks of botulism from fish in the USA in the early 1960s attracted an upsurge of interest which has to some degree persisted since then. A review of botulism was published in 1979[314] and a thorough review of those aspects specifically associated with fishery products in 1976.[315] Because of the environment in which they live, and the fact that fish are poikilothermic animals, botulism from fish is usually caused by psychrotrophic strains. The ecology of these strains of *Cl. botulinum*,[316,317,318] their taxonomy[318] and methods for their isolation and identification[319] have all been well documented and the hazards associated with smoked fish reviewed.[320]

Fish botulism is most often caused by strains producing type E toxin, though the other psychrotrophic strains producing types B and F toxins are frequently associated with fishery products. In their other properties there is little to distinguish the psychrotrophic strains producing any of these toxins. Their minimum growth temperature is 3–5 °C, they are essentially non-proteolytic but actively saccharolytic and lipolytic and are inhibited by 3–5 % salt depending on the temperature. Whilst they are spore-forming bacteria, the spores of these strains are only moderately heat resistant. The available evidence shows that the psychrotrophic strains are widespread in soils in certain parts of the world and in some areas are common in aquatic sediments. Under some circumstances they are able to grow at least in freshwater sediments. They are not normally present in freshly caught fish and shellfish in large numbers and if growth and toxin production does occur in fishery products it is almost always a result of faulty handling and processing at some stage. The toxins themselves are proteins which are released from the vegetative cells when they grow and lyse. They are readily inactivated by heat, are stable at acid pH and unstable at alkaline pH.

The main properties of the organisms and their toxins which are of importance to food processors were known at the time of Van Ermengen's classical report of 1897.[321] Botulism has, therefore, for many years resulted from failure to observe one or other of the known factors which control the growth of the organisms and the toxins they produce. Perhaps the most significant factor to emerge from more recent work is the realisation that the inhibitory effects of parameters such as salt, a_w, pH, temperature and chemical inhibitors are interdependent and there are synergistic effects. This has enabled refinements to be made in specifications for safe foods. An example of this was demonstrated by Smelt[322] who showed that a particular recipe for smoked fish was not sufficient to inhibit type E toxin at ambient temperatures but that inhibition did occur at 8 °C. Another example is lumpfish caviar where a combination of salt and low pH together were responsible for the inhibition of growth of types A, B and E toxins.[323]

Earlier work reviewed by Hobbs[315] showed that different species of fish had varied abilities to support growth and toxin production by Cl. botulinum. Similar results have been found with different recipes for canned fish products.[324,325] Also, earlier work had shown that extracts of various fish differed markedly in their ability to support the germination and outgrowth of spores.[326] These effects were attributed to the concentrations of L-alanine and lactate, both known germinants for Cl. botulinum. The amino acid glycine on the other hand has been shown to inhibit growth and toxin production when added to kamaboko.[327]

Early work on packaging had shown that vacuum packaging had little effect on growth and toxin production and this has been confirmed.[194] The use of packaging in carbon dioxide atmospheres has been investigated[328,329] and shown to have a small inhibitory effect which is probably of itself of little commercial value.

The effect of oxidation–reduction potential (E_h) on growth and toxin production has been investigated[330] and a high E_h together with a lowered pH have been claimed to be two of the more important inhibitory factors in canned fish inoculated with *Cl. botulinum*.[331]

The only practical way of destroying the spores of *Cl. botulinum* in foods is by heat treatment. Prevention of growth, however, can be achieved by refrigeration, freezing, reduction of pH, reduction of water activity (a_w) by the addition of salt or other solutes and in some cases by the addition of chemical preservatives. As has been stated, combinations of two or more of these parameters can be advantageous and many of them can reduce the heat resistance of spores. An unknown water soluble component of vegetable oils has been reported to reduce the heat resistance of spores of a putrefactive anaerobe.[332] Also, the heat resistance of the same organism has been reported to be lower in spoiling fish than in fresh fish.[333]

Irradiation can be used to kill spores but the adverse effects of the doses required on the organoleptic properties of the fish generally preclude its use for sterilisation. Lower doses which extend the shelf life of fishery products cannot be considered to control *Cl. botulinum*.[180]

In raw fishery products the level of contamination with *Cl. botulinum* is low and, if stored at a temperature where growth and toxin production can occur, the fish is normally spoiled before significant amounts of toxin accumulate. Whenever processing is less than sterilisation, spoilage of the product can be delayed sufficiently for edible fish to be toxic. Examples of such processes are heat pasteurisation and smoke curing. In these products other factors such as salt or refrigeration are relied upon to prevent the growth of *Cl. botulinum*. With many there is the additional safeguard that they are cooked before consumption and most normal cooking procedures inactivate the toxins. Others are not cooked and it is these products which represent the greatest hazard.

Low levels of contamination occur with many fresh fish products however, occasionally there are situations where high levels can occur,[315] such as in some fish farms and under unhygienic processing conditions. Although it is unlikely that further research will lead to the total elimination of *Cl. botulinum* from fresh fish, much remains to be understood concerning the conditions which permit the development of high levels of contamination.

In most parts of the world the incidence of *Cl. botulinum* is not sufficiently high to merit continuous monitoring of the fish. Routine testing of this kind is expensive and inefficient, and except where the incidence is known or suspected to be high, it is unlikely to prevent an outbreak. It is generally more efficient and effective to control processing and storage conditions so that contamination, growth and toxin production are prevented.

Scombrotoxin Poisoning

Scombrotoxin poisoning generally involves ingestion of fish from the families *Scomberesocidae* and *Scombridae* which include tuna, bonito and mackerel. The symptoms shown have been of varying severity, but are essentially those of histamine toxicity. This particular hazard is interesting in that while it is believed to result from bacterial growth in the fish no specific pathogenic bacterium has been incriminated.

A thorough review of this subject has been published.[334] Though the cause and precise nature of the toxin are still not known, the factors which give rise to poisoning are reasonably well understood.

Accounts of outbreaks have appeared regularly over the past forty years from Japan, the Pacific Islands, the United States of America and elsewhere. In other parts of the world it appears to be relatively rare. In the UK there had been no accurate accounts of incidents prior to 1978 when a detailed account of one outbreak was published.[335] Since then more than fifty such incidents have been recorded.[336]

A significant number of incidents of scombroid poisoning in the UK in 1979 coincided with an increase in the consumption of mackerel particularly as smoked products. An uncharacteristically large number of incidents occurred in the UK in 1980 though a few of these were associated with non-scombroid fish (unpublished data).

Whilst the precise cause of scombroid poisoning is not known some useful conclusions can be drawn from the available information. It is not a microbial infection but caused by some toxin which accumulates in the fish flesh during storage. The toxin is heat stable, canned fish have caused poisoning and the smoking process appears to have no effect on it. Although the symptoms are those of histamine poisoning, there is ample evidence that scombroid poisoning is not an allergy. Historically the condition has been associated with high levels of histamine in the fish flesh and indeed it was suggested at one time that histamine itself was the toxic factor. Whilst it is generally true that toxic fish have high levels of histamine, this is not always true, nor do fish with high levels always cause poisoning.

In spite of this, histamine levels in the flesh of scombroid fish can be a useful indicator of spoilage. It is clear that some degree of spoilage is necessary for the fish to become toxic and that fish seldom, if ever, become toxic when properly refrigerated at ice temperatures. During storage in melting ice mackerel, for instance, remains edible for about twelve days by which time the histamine concentration is in the region of 3–4 mg/100 g of fish.[337] At elevated temperatures (15–25 °C) high concentrations are rapidly reached and fish can become toxic even though they may still be acceptable to the consumer.

Histamine is formed in the flesh by the action of bacterial decarboxylase enzymes on the amino acid histidine which is present at higher levels in scombroid fish than in most other species. Most bacteria associated with histamine production in fish belong to the family *Enterobacteriaceae* in particular *Proteus morganii*, *Klebsiella pneumoniae* and *Hafnia alvei*.[338–342] This is in accordance with the observations on the effect of storage temperature on histamine production. The bacteria identified as responsible generally have a minimum growth temperature of around 8 °C and grow rapidly above 15 °C.

Notwithstanding the exceptions already referred to, toxic fish generally have histamine levels of more than 20 mg/100 g, and often much higher. In the absence of a specific *in vitro* or animal test for the toxin therefore, histamine determination is the only laboratory test of value for quality control purposes. High levels indicate that the fish has been stored under conditions that could give rise to toxicity. Experience since then has confirmed reports in the literature that effective control of temperature during handling and processing is the main practical means of minimising this hazard.

CONCLUSIONS

In reviewing the microbiological studies on fish handling and processing during the past ten to fifteen years, it is apparent that there has been a shift of emphasis. Although the very earliest bacteriological investigations on seafoods were related to their involvement as carriers of human enteric disease[24] for some time prior to the period under review the emphasis had been on spoilage problems. While these are still important the emphasis now is clearly on questions of hygiene, food poisoning and microbiological specifications. There are good reasons for this change, one of the most important being the increase in international trade in fishery products. This has resulted in a number of investigations directed towards setting microbiological specifications for these products in terms of hygiene and quality. Concurrently, the importance of food hygiene and good hygienic practice in processing and catering has become more generally appreciated.

Microbiological control of processing and specifications for products have, therefore, also become a feature of large scale seafood production.

Probably the biggest drawback to the use of such bacteriological investigations in foods is the length of time required to obtain results. To achieve the aim of improving quality and reducing health risks it is important to obtain the results of analyses quickly. There is already some progress in rapid and automated techniques and advances in methods will undoubtedly lead to speedier results. It is also important not only to obtain a full understanding of spoilage risks but also of the conditions which may permit the growth of dangerous organisms in the foods being investigated.

Apart from irradiation no completely novel methods of food preservation have been proposed during the past fifty years. In retrospect the advantages of freezing and cold-storage, as a means of preserving fish, were relatively slowly exploited initially. However, within the past decade or so large scale processing, efficient cold-chain distribution and a marked increase in the use of supermarket frozen display, catering and domestic freezer cabinets have increased the sales of frozen seafoods in the Western world. The bacteriological problems encountered in traditional fishery products are reasonably well understood as a result of the earlier studies which formed the basis of sound principles for the handling and preservation of them. On an international basis this knowledge can still be applied to developing fisheries and might be extended into the fields of both flavour production and quality in fermented fishery products which are widely used in certain areas.

REFERENCES

1. LOVE, R. M., *The Chemical Biology of Fishes*, 1980, Academic Press, London and New York.
2. SHEWAN, J. M., *Fish as Food*, Vol. I, ed. Bergstrom, G. 1961, Academic Press, London and New York.
3. SHEWAN, J. M., *Recent Advances in Food Science*, eds. Hawthorn, J., and Leitch, J. M. 1962, Butterworths, London.
4. SHEWAN, J. M., *J. appl. Bact.*, 1971, **34**, 299.
5. SHEWAN, J. M., and HOBBS, G., *Prog. Indust. Microbiol.*, 1967, **6**, 169.
6. FIELDS, M. L., RICHMOND, B. S., and BALDWIN, R. E., *Adv. Food Res.*, 1968, **16**, 161.
7. CONNELL, J. J., and SHEWAN, J. M., *Advances in Fish Science and Technology*, ed. Connell, J. J. 1980, Fishing News Books Ltd, Farnham, Surrey, England.
8. MORITA, R. Y., *Bact. Rev.*, 1975, **39**, 144.

9. SHEWAN, J. M., HOBBS, G., and HODGKISS, W., *J. appl. Bact.*, 1960, **23**, 379.
10. Symposium on *Pseudomonas* and *Achromobacter*, *J. appl. Bact.*, 1960, **23**, 373.
11. INGRAM, M., and SHEWAN, J. M., *J. appl. Bact.*, 1960, **23**, 373.
12. GIBSON, D. M., HENDRIE, M. S., HOUSTON, N. C. and HOBBS, G., *Aquatic Microbiology*, eds. Skinner, F. S., and Shewan, J. M. 1977, Academic Press, London and New York.
13. BREED, R. S., MURRAY, E. G. D., and SMITH, N. R., *Bergey's Manual of Determinative Bacteriology*, 7th Edn, 1957, Bailliere, Tindall & Cox Ltd, London.
14. BREED, R. S., MURRAY, E. G. D., and HITCHENS, A. P., *Bergey's Manual of Determinative Bacteriology*, 6th Edn, 1948, Williams & Wilkins, Co., Baltimore.
15. BAUMANN, P., DOUDOROFF, M., and STANIER, R. Y., *J. Bact.*, 1968, **95**, 58.
16. BAUMANN, P., DOUDOROFF, M., and STANIER, R. Y., *J. Bact.*, 1968, **95**, 1520.
17. HENDRIE, M. S., HOLDING, A. J., and SHEWAN, J. M., *Int. J. Syst. Bact.*, 1974, **24**, 534.
18. BUCHANAN, R. E., and GIBBONS, N. E. (eds.), *Bergey's Manual of Determinative Bacteriology*, 8th Edn, 1974, Williams & Wilkins, Co., Baltimore.
19. BAUMANN, L., BAUMANN, P., MANDEL, M., and ALLEN, R. D., *J. Bact.*, 1972, **110**, 402.
20. LEE, J. V., GIBSON, D. M., and SHEWAN, J. M., *J. gen. Microbiol.*, 1977, **98**, 439.
21. GRAY, P. A., and STEWART, D. J., *J. appl. Bact.*, 1980, **49**, 375.
22. HORSLEY, R. W., *J. Fish Biol.*, 1977, **10**, 529.
23. SHEWAN, J. M., *Handling, Processing and Marketing of Tropical Fish*, 1977, Tropical Products Institute, London.
24. LISTON, J., Advances in Fish Science and Technology, eds. Connell, J. J., et al., 1980, Fishing News Books Ltd, Farnham, Surrey.
25. HERBORG, L., and VILLADSEN, A., *J. Food Technol.*, 1975, **10**, 507.
26. CANTONI, C., CATTANEO, P. and AUBERT, S. D., *Ind. Aliment.*, 1976, **15**(5), 105.
27. WYATT, L. E., *Diss. Abstr. Int. B.*, 1978, **38**(11), 5206.
28. NAIR, R. B., THORAMANI, P. K., and LAHIRY, N. L., *J. Food Sci. Technol.* (Mysore), 1971, **8**(2), 53.
29. HODGKISS, W., *Hygienic Design and Operation of Food Plant*, ed. Jowitt, R., 1980, Ellis Horwood Ltd, Chichester, England.
30. SHEWAN, J. M., HOBBS, G., and HODGKISS, W., *J. appl. Bact.*, 1960, **23**, 463.
31. GILLESPIE, W. C., and MACRAE, I. C., *J. appl. Bact.*, 1975, **39**, 91.
32. LISTON, J., *J. gen. Microbiol.*, 1957, **16**, 205.
33. COLWELL, R. R., and LISTON, J., *Pacific Sci.*, 1962, **16**, 264.
34. KARTHIAYANI, T. C., and IYER, K. M., *Fish Technol.*, 1967, **4**, 89–97.
35. SIMIDU, U., KANEKO, E., and AISO, K., *Bull. Jap. Soc. Sci. Fish.*, 1969, **35**, 77.
36. OLLEY, J., and RATKOWSKY, D. A., *Food Technol. Austral.*, 1973, **25**(2), 66.
37. CLARK, D. S., *Can. J. Microbiol.*, 1971, **17**, 943.
38. BOUSFIELD, I. J., SMITH, G. L., and TRUEMAN, R. W., *J. appl. Bact.*, 1973, **36**, 297.

39. SHARPE, A. N., and JACKSON, A. K., *Appl. Microbiol.*, 1972, **24**, 175.
40. SIMIDU, U., and HASUO, K., *J. gen. Microbiol.*, 1968, **52**, 347.
41. SIMIDU, U., and HASUO, K., *J. gen. Microbiol.*, 1968, **52**, 355.
42. LEE, J. S., and PFEIFER, D. K., *J. Milk Food Technol.*, 1974, **37**, 553.
43. SHARPE, A. N., and CLARK, D. S. (eds.), *Mechanizing Microbiology*, 1978, C. C. Thomas, Springfield, Illinois.
44. SHARPE, A. N., *Sampling—Microbiological Monitoring of Environments*, Society for Applied Bacteriology, Technical Series No. 7. eds. Board, R. G., and Lovelock, D. W., 1973, Academic Press, London and New York.
45. JARVIS, B., LACH, V. H. and WOOD, J. M., *J. appl. Bact.*, 1977, **43**, 149.
46. KRAMER, J. M., KENDALL, M., and GILBERT, R. J., *Eur. J. appl. Microbiol. Biotechnol.*, 1979, **6**, 289.
47. DONNELLY, C. B., GILCHRIST, J. E., PEELER, J. T., and CAMPBELL, J. E., *Appl. Environ. Microbiol.*, 1976, **32**, 21.
48. HERBERT, R. A., HENDRIE, M. S., GIBSON, D. M., and SHEWAN, J. M., *J. gen. Microbiol.*, 1971, **44**, 419.
49. LEVIN, R. E., *Appl. Microbiol.*, 1968, **16**, 1734.
50. SADOVSKI, E. Y., and LEVIN, R. E., *Appl. Microbiol.*, 1969, **17**, 787.
51. CHAI, T., CHEN, C., ROSEN, A., and LEVIN, R. E., *Appl. Microbiol.*, 1968, **16**, 1738.
52. CHEN, T. C., NAWAR, W. W., and LEVIN, R. E., *Appl. Microbiol.*, 1974, **28**, 679.
53. CHEN, T. C., and LEVIN, R. E., *Appl. Microbiol.*, 1974, **28**, 681.
54. LERKE, P., FARBER, L., and ADAMS, R., *Appl. Microbiol.*, 1967, **15**, 77.
55. SHAW, B. G., and SHEWAN, J. M., *J. appl. Bact.*, 1968, **31**, 89.
56. HERBERT, R. A., and HAIGHT, R. D., *J. appl. Bact.*, 1967, **30**, 224.
57. HERBERT, R. A., ELLIS, J. R., and SHEWAN, J. M., *J. Sci. Fd Agric.*, 1975, **26**, 1187.
58. HERBERT, R. A., and SHEWAN, J. M., *J. Sci. Fd Agric.*, 1976, **27**, 89.
59. MILLER, A., SCANLAN, R. A., LEE, J. S., and LIBBEY, L. M., *J. agric. Fd Chem.*, 1972, **20**, 709.
60. MILLER, A., SCANLAN, R. A., LEE, J. S. and LIBBEY, L. M., *Appl. Microbiol.*, 1973, **25**, 257.
61. MILLER, A., SCANLAN, R. A., LEE, J. S., and LIBBEY, L. M., *Appl. Microbiol.*, 1973, **25**, 952.
62. MILLER, A., SCANLAN, R. A., LEE, J. S., and LIBBEY, L. M., *Appl. Microbiol.*, 1973, **26**, 18.
63. LAYCOCK, R. A., and REGIER, L. W., *J. Fish Res. Bd Can.*, 1971, **28**, 305.
64. EASTER, M. C., GIBSON, D. M., and WARD, F. B., *J. appl. Bact.*, in press.
65. VAN SPRECKENS, K. J. A., *Antonie van Leeuwenhoek*, 1977, **43**, 283.
66. SHEWAN, J. M., and MURRAY, C. K., *Cold Tolerant Microbes in Spoilage and the Environment*, eds. Russell, A. D., and Fuller, R. 1979, Academic Press, London and New York.
67. CANN, D. C., *Handling, Processing Marketing of Tropical Fish*, 1977, Tropical Products Institute, London.
68. MURRAY, C. K. and FLETCHER, T. C., *J. Fish Biol.*, 1976, **9**, 329.
69. CANN, D. C., *Fishery Products*, ed. Kreuzer, R., 1974, Fishing News Books Ltd, Farnham, Surrey, England.

70. LEE, J. S. and PFEIFER, D. K., *Appl. Microbiol.*, 1975, **30**, 72.
71. LEE, J. S., and PFEIFER, D. K., *Appl. Environ. Microbiol.*, 1977, **33**, 853.
72. KOBURGER, J. A., NORDEN, A. R., and KEMPLER, G. M., *J. Milk Food Technol.*, 1975, **38**, 747.
73. MURCHELAW, R. A., and BISHOP, J. L., *J. Invert. Path.*, 1969, **14**, 321.
74. VANDERZANT, C., MROZ, E., and NICKELSON, R., *J. Milk Food Technol.*, 1970, **33**, 346.
75. VANDERZANT, C., NICKELSON, R., and JUDKINS, P. W., *Appl. Microbiol.*, 1971, **21**, 916.
76. VANDERZANT, C., JUDKINS, P. W., NICKELSON, R., and FITZHUGH, H. A., *Appl. Microbiol.*, 1972, **23**, 38.
77. VANDERZANT, C., COBB, B. T., THOMPSON, C. A., and PARKER, J. C., *J. Milk Fd Technol.*, 1973, **36**, 443.
78. CHRISTOPHER, F. M., VANDERZANT, C., PARKER, J. D., and CONTE, F. S., *J. Food Prot.*, 1978, **41**, 20.
79. COX, N. A., and LOVELL, R. T., *J. Food Sci.*, 1973, **38**, 679.
80. WEBB, N. B., and THOMAS, F. B., *N. Carolina Dept. Conservation and Devel. Sci. Rpt. No. 16*, 1968.
81. VASCONCELOS, G. J., and LEE, J. S., *Appl. Microbiol.*, 1972, **23**, 11.
82. MÜNZNER, R., *Arch. Lebensmittelhyg.*, 1976, **27** (4), 136.
83. RAY, B., WEBB, N. B., and SPECK, M. L., *J. Food Sci.*, 1976, **41**, 398.
84. WARD, D. R., PIERSON, M. D., and VAN TASSELL, K. R., *J. Food Sci.*, 1977, **42**, 597.
85. PHILLIPS, F. A., and PEELER, J. T., *Appl. Microbiol.*, 1972, **24**, 958.
86. LOAHARANN, P., and LOPEZ, A., *Appl. Microbiol.*, 1970, **19**, 734.
87. WARD, J. F., and TATRO, M. C., *Chesapeake Sci.*, 1971, **11** (3), 193.
88. TANIKAWA, E., *Marine Products in Japan*, 1971, Kosheisha-Koseikaku Co., Tokyo.
89. ISHII, S., and AMANO, K., *Fishery Products*, ed. Kreuzer, R., 1974, Fishing News Books Ltd, Farnham, Surrey, England.
90. BOND, R. M., *FAO Fish Circ.*, 1975, No. 332, FAO, Rome.
91. KEAY, J. N. (ed.), *The Production and Utilisation of Mechanically Recovered Fish Flesh (Minced Fish)*, ed. Keay, J. N., 1976, Torry Research Station, Aberdeen.
92. BLACKWOOD, C. M., *Fishery Products*, ed. Kreuzer, R., 1974, Fishing News Books Ltd, Farnham, Surrey, England.
93. CRABB, W. E., and GRIFFITHS, D. J., *The Production and Utilisation of Mechanically Recovered Fish Flesh (Minced Fish)*, ed. Keay, J. N., 1976, Torry Research Station, Aberdeen.
94. The International Commission on Microbiological Specifications for Foods (ICMSF), *Micro-Organisms in Foods* Vol. 2, 1974, University of Toronto Press, Toronto and Buffalo, Canada.
95. LICCIARDELLO, J. J., and HILL, W. S., *J. Food. Prot.*, 1978, **41**, 948.
96. CANN, D. C., and TAYLOR, L. Y., *The Production and Utilisation of Mechanically Recovered Fish Flesh (Minced Fish)*, ed. Keay, J. N., Torry Research Station, Aberdeen.
97. RACCACH, M., and BAKER, R. C., *J. Food Sci.*, 1978, **43**, 1675.
98. MOTEGI, S., *Bull. Jap. Soc. Sci. Fish.*, 1979, **45**, 79.

99. MORI, K., SAWADA, H., NABETANI, O., MARUO, S. and HIRANO, T., *Bull. Jap. Soc. Sci. Fish.*, 1973, **39**, 1063.
100. MORI, K., NABETANI, O., and HIRANO, T., *Bull. Jap. Soc. Sci. Fish.*, 1974, **40**, 959.
101. FUJITA, Y., and MIYAMOTO, M., *Bull. Jap. Soc. Sci. Fish.*, 1975, **41**, 1263.
102. FUJITA, Y., MIYAMOTO, M., and MATSUDA, T., *Bull. Jap. Soc. Sci. Fish.*, 1974, **40**, 825.
103. KING, F. S., TANG, N. Y., and CAVALLETTO, C., *J. Food Sci.*, 1972, **37**, 191.
104. SAKAGUCHI, G., *Food-Borne Infections and Intoxications*, ed. Riemann, H., 1969, Academic Press, London and New York.
105. DALEY, L. H., DENG, J. C., and OBLINGER, J. L., *J. Food Sci.*, 1979, **44**, 883.
106. STRØM, A. R., and LARSEN, H., *J. appl. Bact.*, 1979, **46**, 531.
107. STRØM, A. R., OLAFSEN, J. A., REFSUES, K. H., and LARSEN, H., *J. appl. Bact.*, 1979, **46**, 545.
108. SCHRØDER, K., CLAUSEN, E., SANDBERG, A. M., and RAA, J., *Advances in Fish Science and Technology*, eds. Connell, J. J., *et al.*, 1980, Fishing News Books Ltd, Farnham, Surrey, England.
109. KRAUS, H., *Arch. Lebensmittelhyg.*, 1961, **12**, 101.
110. LINDGREN, S., and CLEVESTRÖM, G., *Swedish J. Agric. Res.*, 1978, **8**, 61.
111. INGRAM, M., and MACKEY, B. M., *Inhibition and Inactivation of Vegetative Microbes*, Soc. Appl. Bact. Symp. Series No. 5, 1975, Academic Press, London and New York.
112. MACLEOD, R. A. and CALCOTT, P. H., *The Survival of Vegetative Microbes*, eds. Gray, T. R. G., and Postgate, J. R., 1976, Cambridge University Press, London, New York and Melbourne.
113. CALCOTT, P. H., *Freezing and Thawing Microbes*, 1978, Meadowfield Press, Ltd, Bushey, England.
114. STRAKA, R. P., and STOKES, J. L., *J. Bact.*, 1959, **78**, 181.
115. HALL, L. P., *A Manual of Methods for the Bacteriological Examination of Frozen Foods*, 1975, Oxford University Press, Oxford.
116. CASEVIO, G., TIREO, G., and GENNARI, M., *Archo. vet. ital.*, 1975, **26**, 121.
117. RAY, B., *J. Food Prot.*, 1979, **42**, 346.
118. SPLITTSTOESSER, D. F., *Food Microbiology: Public Health and Spoilage Aspects*, eds. de Figueiredo, M. P., and Splittstoesser, D. F., 1976, Avi Publishing Co., Westport, Conn.
119. SPECK, M. L., and RAY, B., *J. Food Prot.*, 1977, **40**, 333.
120. KRAFT, A. A., and REY, C. R., *Food Technol.*, 1979, **33** (1), 66.
121. MICHENER, H. D., *Food Microbiology and Technology*, eds. Jarvis, B., Christian, J. H. B., and Michener, H. D., 1979, Medicina Viva Servicio Congressi, SrI. Parma, Italy.
122. MOSSEL, D. A. A., VELDMAN, E., and EELDERINK, I., *J. appl. Bact.*, 1980, **49**, 405.
123. FLOWERS, R. S., MARTIN, S. E., BREWER, D. G., and ORDAL, Z. J., *Appl. Environ. Microbiol.*, 1977, **33**, 1112.
124. CAMPELLO, F., *Revue gen. Froid Ind. frigor*, 1970, **61**, 1445.
125. GJERDE, J., *Fisk. Ski. Ser. Ernaering*, 1976, **1**, 17.
126. GJERDE, J., *Svensk. Vet.*, 1976, **28**, 911.

127. BAER, E. F., DURAN, A. D., LEININGOR, H. V., READ, R. B., SCHWAB, R. H., and SWARTZENTRUBER, A., *Appl. Environ. Microbiol.*, 1976, **31**, 337.
128. HAYWARD, M. J., and MacCALLUM, W. A., *J. Fish. Res. Bd Can.*, 1969, **26**, 3217.
129. HAYWARD, M. J., MacCALLUM, W. A., and SHAW, D. H., *J. Fish Res. Bd Can.*, 1970, **27**, 1983.
130. VENKATARAMAIAH, N., and KEMPTON, A. G., *Can. J. Microbiol.*, 1975, **21**, 1788.
131. MACKIE, I. M., HARDY, R., and HOBBS, G., *FAO Fishery Rpt*, 1971, No. 100, FAO Rome.
132. BARTL, V., The Microbiological Safety of Foods, eds. Hobbs, B. C., and Christian, J. H. B., 1973, Academic Press, London and New York.
133. HAMED, M. G., ELIAS, A. N., EL-WAKEIL, F. A., and FODA, I. O., *Egyptian J. Food Sci.*, 1973, **1**, 1.
134. FUJII, T., ISHIDA, Y., and KADOTA, H., *Bull. Jap. Soc. Sci. Fish.*, 1977, **43**, 1241.
135. FUJII, T., *Bull. Jap. Soc. Sci. Fish.*, 1977, **43**, 609.
136. COLWELL, R. R., LITCHFIELD, C. D., VREELAND, R. H., GIBBONS, N. E., and KIEFER, L. A., *Int. J. Syst. Bact.*, 1979, **29**, 379.
137. KOVALEVA, N. A., KONNOVA, A. A., TSAPLINA, I. A., and GORBUNOV, K. V., *Prikl. Biokhem. Mikrobiol.*, 1977, **13**, 501.
138. GONZALEZ, C., and GUTIERREZ, C., *Microbiologia esp.*, 1970, **23**, 223.
139. DEL VALLE, F. R., and GONZALES-INIGO, J. L., *Food Technol.*, 1968, **22**, 1135.
140. DEL VALLE, F. R., and NICKERSON, J. T. R., *Food Technol.*, 1968, **22**, 1036.
141. DEL VALLE, F. R., BOURGES, H., HAAS, R., and GAONA, H., *J. Food Sci.*, 1976, **41**, 975.
142. WOJTOWICZ, M. B., FIERHELLER, M. G., LEGENDRE, R., and REGIER, L. W., *Tech. Rep. Fish & Mar. Serv. (Can)*, 1977, No. 731.
143. VARGA, S., SIMS, G. G., MICHALIK, P., and REGIER, L. W., *J. Food Sci.*, 1979, **44**, 47.
144. DOFSBURG, J. J., LAMPRECHT, F. C., ELLIOT, M. C., and REID, D. A., *J. Food Technol.*, 1969, **4**, 339.
145. FONG, Y. Y., and WALSH, E. O'F., *Lancet*, 1971, **II** (7732), 1032.
146. FONG, Y. Y., and CHAN, W. C., *Nature (Lond.)*, 1973, **243** (5407) 421.
147. HANSEN, P., *Neue Verpack.*, 1972, **25**, 1164.
148. LEE, K. H., *J. Korean Agric. Chem. Soc.*, 1969, **11**, 1.
149. ORILLO, C. A., and PEDERSON, C. S., *Appl. Microbiol.*, 1968, **16**, 1669.
150. SANDS, A., and CRISAN, F. V., *J. Food Sci.*, 1974, **39**, 1002.
151. LEE, J. S., and PFEIFER, D. K., *J. Milk Fd Technol.*, 1973, **36**, 143.
152. DENG, J., TOLEDO, R. T., and LILLARD, D. A., *J. Food Sci.*, 1974, **39**, 596.
153. TEPLITSKAYA, A. M., MYAKISHEVA, A. K., and RESHETNYAK, M. S., *Ryb. Khoz.*, 1969, **45** (4), 67.
154. MILER, K. B. M., and KOZLOWSKI, Z. P., *Rev. Conserve*, 1971, **28**, 122.
155. OLSEN, C. Z., *Proc. Europ. Meeting Meat Res. Workers*, 1976, No. 22, F7-1, F7-10.
156. HOBBS, G., and SHEWAN, J. M., *Freezing and Irradiation of Fish*, ed. Kreuzer, R., 1969, Fishing News Books Ltd, Farnham, Surrey, England.

157. International Atomic Energy Agency and FAO, *Food Irradiation*, 1966, IAEA, Vienna.
158. KREUZER, R. (ed.), *Freezing and Irradiation of Fish*, 1969, Fishing News Books Ltd, Farnham, Surrey, England.
159. International Atomic Energy Agency and FAO, *Food Irradiation*, 1973, IAEA, Vienna.
160. International Atomic Energy Agency and FAO, *Application of Food Irradiation in Developing Countries*, 1966, IAEA, Vienna.
161. International Atomic Energy Agency, *Microbiological Problems in Food Preservation by Irradiation*, 1967, IAEA, Vienna.
162. International Atomic Energy Agency and FAO, *Elimination of Harmful Organisms from Food and Feed by Irradiation*, 1968, IAEA, Vienna.
163. International Atomic Energy Agency and FAO, *Preservation of Fish by Irradiation*, 1970, IAEA, Vienna.
164. LEE, J. S., WILLETT, C. L., ROBINSON, S. M., and SINNHABER, R. O., *Appl. Microbiol.*, 1967, **15**, 368.
165. LICCIARDELLO, J. J., RONSIVALLI, L. J., and SLAVIN, J. W., *J. appl. Bact.*, 1967, **30**, 239.
166. KAWABATA, T., KOZIMA, T., and OKITSU, T., *Food Irradiation (Shokuhin-Shosha)*, 1968, **3**, 40.
167. LAYCOCK, R. A., and REGIER, L. H., *Appl. Microbiol.*, 1970, **20**, 333.
168. MCKINNEY, F. E., *Isotopes Rad. Technol.*, 1971, **9**, 75.
169. KUMTA, U. S., and MAVINKURVE, S. S., *J. Food Sci.*, 1971, **36**, 63.
170. MIYAUCHI, D., *Isotope Rad. Technol.*, 1972, **9**, 299.
171. HANNESSON, G., *Inds. aliment agric.*, 1973, **89**, 205.
172. SASAYAMA, S., *Bull. Tokai Reg. Fish. Res. Lab.*, 1973, No. 75, 39.
173. DOKE, S. N., GHADI, S. V., ALUR, M. D., KUMTA, U. S., and LEWIS, N. F., *Acta Aliment.*, 1975, **5**, 139.
174. KARNOP, G., *Chem. Mikrobiol. Technol Lebensm*, 1975, **4** (2), 40.
175. BANIK, A. K., CHAUDURI, D. R., and BOSE, A. N., *J. Food Sci. Technol. India*, 1976, **13**, (2), 67.
176. BANIK, A. K., CHAUDURI, D. R., and BOSE, A. N., *J. Food Sci. Technol. India*, 1976, **13** (2), 72.
177. VAN SPREEKENS, K. J. A., and TOEPOEL, L., *Food Preservation by Irradiation*, Vol. II, International Atomic Energy Agency, FAO, 1978, IAEA, Vienna.
178. HUSSAN, A. M., EHLERMANN, D., and DIEHL, J. F., *Arch Lebensmittelhyg.*, 1976, **27**, 223.
179. HANNESSON, G., and DAGBJARTSSON, B., *Isotopes Rad. Technol.*, 1972, **9**, 468.
180. HOBBS, G., *Food Irradiation Information No. 7*, 1977, International Project in the Field of Food Irradiation, Karlsruhe.
181. KRABBENHOFT, K. L., CORLETT, D. A., ANDERSON, A. W., and ELLIKER, P. R., *Appl. Microbiol.*, 1964, **12**, 424.
182. GRECZ, N., UPADHYAY, J., TANG, T. C., and LIN, C. A., *Microbiological Problems in Food Preservation by Irradiation*, IAEA Panel Proceedings Series, 1967, IAEA, Vienna.
183. DURBAN, E., and GRECZ, N., *Appl. Microbiol.*, 1969, **18**, 44.
184. DURBAN, E., GROODNOW, R., and GRECZ, N., *J. Bact.*, 1970, **102**, 590.
185. HARDY, R., and HOBBS, G., *Plastics & Polymers*, 1968, **36**, 445.

186. SHEWAN, J. M., and HOBBS, G., *Fishing News Int.*, 1963, **2** (1), 103.
187. DEBEVERE, J. M., and VOETS, J. P., *J. appl. Bact.*, 1971, **34**, 507.
188. DEBEVERE, J. M., and VOETS, J. P., *J. appl. Bact.*, 1972, **35**, 351.
189. HANSEN, P., *J. Food Technol.*, 1972, **7**, 21.
190. HUSS, H. H., *J. Food Technol.*, 1972, **7**, 13.
191. DEBEVERE, J. M., and VOETS, J. P., *Lebensm.-Wiss. Technol.*, 1974, **7** (2), 73.
192. ABRAHAMSON, K., DE SILVA, W. N., and MOLIN, N., *Can. J. Microbiol.*, 1965, **11**, 523.
193. CANN, D. C., WILSON, B. B., HOBBS, G., and SHEWAN, J. M., *J. appl. Bact.*, 1965, **28**, 431.
194. CANN, D. C., TAYLOR, L. Y., and COLLETT, J. M., *Proceedings World Congress Foodborne Infections and Intoxications*, West Berlin, 1980, in press.
195. ENFORS, S.-O., MOLIN, G., and TERNSTRÖM, A., *J. appl. Bact.*, 1979, **47**, 197.
196. GILL, C. O., and TAN, K. H., *Appl. Environ. Microbiol.*, 1980, **39**, 317.
197. ENFORS, S.-O., and MOLIN, G., *J. appl. Bact.*, 1980, **48**, 409.
198. HUSS, H. H., SCHAEFFER, I., PEDERSEN, A., and JEPSON, A., *Advances in Fish Science and Technology*, eds. Connell, J. J. *et al.*, 1980, Fishing News Books Ltd, Farnham, Surrey, England.
199. HOBBS, G., *IFST Proc.*, 1970, **3**, 98.
200. WUTHE, H. H., and FINDEL, G., *Arch. Lebensmittelhyg.*, 1972, **23** (5), 110.
201. BUROW, H., *Arch. Lebensmittelhyg.*, 1974, **25** (2), 39.
202. MATCHES, J. R., and LISTON, U., *J. Milk Food Technol.*, 1972, **35**, 39.
203. STEPHEN, S., INDRANI, R., KOTIAN, M., and RAO, K. N. A., *Indian J. Microbiol.*, 1975, **15** (2), 64.
204. PANDURANGA RAO, C. C., and GUPTA, S. S., *Fish Technol.*, 1978, **15**, 45.
205. WYATT, L. E., NICKELSON, R., and VANDERZANT, C., *J. Food Sci.*, 1979, **44**, 1067.
206. RAJ, H. D., *Lab. Pract.*, 1979, **19**, 374, 394.
207. SUMNER, J. L., BOYD, S., and WILSON, N. D., *Commercial Fishing (NZ)*, 1972 (Nov.), 15.
208. WEBB, W. B., STOKES, S. J., THOMAS, F. B., MONCOL, N. B., and HARDY, E. R., *Proc. Gulf. Carib. Fish Inst.*, 1973 (May), 109.
209. ROBE, K., *Food Process.*, 1974, **35** (2), 19.
210. PHILLIPS, W. K., and HOLLIS, C. G., *Dev. Ind. Bact.*, 1975, **16**, 490.
211. HEDÉN, C-G., and ILLÉNI, T., *New Approaches to the Identification of Microorganisms*, 1974, John Wiley & Sons, London.
212. HEDÉN, C-G., and ILLÉNI, T., *Automation in Microbiology and Immunology*, 1975, John Wiley & Sons, London.
213. JOHNSTON, H. H., and NEWSOM, S. W. B. (eds.), *Rapid Methods and Automation in Microbiology*, 1976, Learned Information, Oxford.
214. ANDERSON, J. M., and BAIRD-PARKER, A. C., *J. appl. Bact.*, 1975, **39**, 111.
215. ANDREWS, W. H., DIGGS, C. D., and WILSON, C. R., *Appl. Microbiol.*, 1975, **29**, 130.
216. FRANCIS, D. W., and TWEDT, R. M., *Abstr. Ann. Meeting, Amer. Soc. Microbiol.*, 1975, **75**, 201.
217. QUADRI, R. B., BUCKLE, K. A., and EDWARDS, R. A., *J. appl. Bact.*, 1974, **37**, 7.
218. ZALESKI, S., and FIK, A., *Medycyna wet.*, 1975, **31**, 631.

219. GODWIN, G. J., GRODNER, R. M., and NOVAK, A. F., *J. Food Sci.*, 1977, **42**, 750.
220. INSALATA, N. F., MAHNKE, C. W., DUNLAP, W. G., and BEAZLEY, C. C., *Food Technol.*, 1972, **26** (5), 78.
221. SHAHIOI, S. A., and FERGUSON, A. R., *Appl. Microbiol.*, 1971, **21**, 500.
222. KANZAKI, M., JINBO, K., MURAKAMI, H., and HARUTA, M., *Ann Rept. Tokyo Metropol. Res. Lab. Publ. Hlth.*, 1975, **26**, 168.
223. CHANG, D. S., and CHOE, W. K., *Bull. Korean Fish Soc.*, 1973, **6** (3/4), 92.
224. ANDREWS, W. H., DIGGS, C. D., PRESNELL, M. W., MIESCIER, J. J., WILSON, C. R., GOODWIN, C. P., ADAMS, W. N., FURFARI, S. A., and MUSSELMAN, J. F., *Abstr. Ann. Meeting. Amer. Soc. Microbiol.*, 1974, **74**, 65.
225. AYRES, P. A., *J. Hyg. (Camb.)*, 1975, **4**, 431.
226. DAVIS, J. G., *Lab. Pract.*, 1969, **18**, 749.
227. DAVIS, J. G., *Lab. Pract.*, 1969, **18**, 839.
228. SHEWAN, J. M., *Chem. Ind.*, 1970, **6**, 193.
229. KEMPA, W., *J. Milk Food Technol.*, 1973, **36**, 392.
230. CORLETT, D. A., *Food Technol.*, 1974, **28** (10), 34.
231. FOSTER, E. M., *Bull. Assoc. Food Drug. Off. US*, 1974, **38**, 267.
232. YETERIAN, M., CHUGG, L., SMITH, W., and COLES, C., *Food Technol.*, 1974, **28** (10), 23.
233. JARVIS, B., *J. Assoc. Publ. Anal.*, 1976, **14** (3), 91.
234. LECHOVICH, R. V., *Ass. Fd Drug. Off. Quart. Bull.*, 1974, **40** (1), 27.
235. ROBERTS, T. A., *Inst. Meat Bull.*, 1976, **94**, 24.
236. SHEWAN, J. M., *Food Technol. Austral.*, 1976, **28**, 493.
237. SPARNON, R. B., *J. Assoc. Publ. Anal.*, 1976, **14** (3), 87.
238. CHARLES, R. H. G., *Health Trends*, 1979, **11** (1), 1.
239. International Commission on Microbiological Specifications for Foods, *Micro-Organisms in Foods*, Vol. 1, 1978, University of Toronto Press, Toronto.
240. FUJINO, T., *International Symposium on* Vibrio parahaemolyticus, eds. Fujino, T., Sakaguchi, G., Sakazaki, R., and Takeda, Y., 1974, Saikon Publishing Co., Tokyo.
241. FUJINO, T., SAKAGUCHI, R., SAKAZAKI, R., and TAKEDA, Y. (eds.), *International Symposium on* Vibrio parahaemolyticus, 1974, Saikon Publishing Co., Tokyo.
242. BAROSS, J., and LISTON, J., *Nature (Lond.)*, 1968, **217**, 1263.
243. BARKER, W. H., WEAVER, R. E., MORRIS, G. K., and MARTIN, W. T., *Microbiology—1974*, ed. Schlessinger, D., 1975, American Society for Microbiology, Washington, DC.
244. COLWELL, R. R., *Microbiology—1974*, ed. Schlessinger, D., 1975, American Society for Microbiology, Washington, DC.
245. TAKIKAWA, I., *Yokohama Med. Bull.*, 1958, **9**, 313.
246. MIYAMOTO, Y., NAKAMURA, K., and TAKIZAWA, K., *Jap. J. Microbiol.*, 1961, **5**, 477.
247. SAKAZAKI, R., IWANAMI, S., and FUKUMI, H., *Jap. J. Med. Sci. Biol.*, 1963, **16**, 161.
248. ZEN-YOJI, H., SAKAI, S., TERAYAMA, T., KUDO, Y., ITO, T., BENOKI, M., and NAGASAKI, M., *J. Infect. Dis.*, 1965, **115**, 436.

249. FUJINO, T., MIWATANI, T., KONDO, M., TAKEDA, Y., AKITA, Y., KOTERA, K., OKADA, H., NISHIMUNE, H., SHIMIZU, Y., TAMURA, T., and TAMURA, Y., *Biken J.*, 1966, **9**, 215.
250. SAKAZAKI, R., *Jap. J. Med. Sci. Biol.*, 1968, **21**, 359.
251. COLWELL, R. R., *J. Bact.*, 1970, **104**, 410.
252. BAUMANN, P., BAUMANN, L., and MANDEL, M., *J. Bact.*, 1971, **107**, 268.
253. ALLEN, R. D., and BAUMANN, P., *J. Bact.*, 1971, **107**, 295.
254. BAUMANN, P., and BAUMANN, L., *Ann. Rev. Microbiol.*, 1977, **31**, 39.
255. BAUMANN, P., BAUMANN, L., BANG, S. S., and WOODKALIS, M. J., *Curr. Microbiol.*, 1980, **4**, 127.
256. International Committee on Systematic Bacteriology, Sub-committee on the taxonomy of vibrios. *Int. J. Syst. Bact.*, 1975, **25**, 389.
257. HORIE, S., OKUZUMI, M., KATO, N., and SAITO, K., *Bull. Jap. Soc. Sci. Fish.*, 1966, **32**, 424.
258. TWEDT, R. M., SPAULDING, P. L., and HALL, H. E., *J. Bact.*, 1969, **98**, 511.
259. BAROSS, J., and LISTON, J., *Appl. Microbiol.*, 1970, **20**, 179.
260. LEE, S. J., *Appl. Microbiol.*, 1972, **23**, 166.
261. KATOH, H., *Jap. J. Bact.*, 1965, **20**, 94.
262. BARROW, G. I., and MILLER, D. C., International Symposium on *Vibrio parahaemolyticus*, eds. Fujino, T. *et al.*, 1974, Saikon Publishing Co., Tokyo.
263. BEUCHAT, L. R., *J. Milk Food Technol.*, 1975, **38**, 476.
264. HENDRIE, M. S., HODGKISS, W., and SHEWAN, J. M., *J. gen. Microbiol.*, 1970, **64**, 151.
265. FURNISS, A. L., LEE, J. V., and DONOVAN, T. J., *The Vibrios*, 1978, PHLS Monograph Series, HMSO, London.
266. BARROW, G. I., and MILLER, D. C., *Microbiology in Agriculture, Fisheries and Food*, eds. Skinner, F. A., and Carr, J. G., 1976, Academic Press, London and New York.
267. NAIR, G. B., ABRAHAM, M., and NATARAJAN, R., *Can. J. Microbiol.*, 1980, **26**, 1264.
268. SAKAZAKI, R., *Food-borne Infections and Intoxications*, ed. Riemann, H., 1969, Academic Press, London and New York.
269. NICKELSON, R., and VANDERZANT, C., *J. Milk Food Technol.*, 1971, **34**, 447.
270. THOMSON, W. K. and THACKER, C. L., *J. Fish. Res. Bd Can.*, 1972, **29**, 1633.
271. LISTON, J., *Microbial Safety of Fishery Products*, eds. Chichester, C. O., and Graham, H. D., 1973, Academic Press, London and New York.
272. LISTON, J., and BAROSS, J., *J. Milk Food Technol.*, 1973, **36**, 113.
273. BARROW, G. I., *Microbiological Safety of Foods*, eds. Hobbs, B. C., and Christian, J. H. B., 1973, Academic Press, London and New York.
274. PAN-URAI, R., BURKHARDT, F., and SAMAHAN, A., *Zbl. Bakt. Hyg.*, Abt. I, Orig. A., 1973, **225**, 46.
275. BARROW, G. I., *Postgrad. Med. J.*, 1974, **50**, 612.
276. CABASSI, E., and MORI, L., *Folia Vet. Lat.*, 1976, **6**, 335.
277. MIWATANI, T., and TAKEDA, Y., Vibrio parahaemolyticus, *a Causative Organism of Food Poisoning*, 1976, Saikon Publishing Co., Tokyo.
278. RICHARD, C., and LHUILLIER, M., *Bull. Inst. Past.*, 1977, **75**, 345.
279. AYRES, P. A., and BARROW, G. I., *J. Hyg. (Camb.)*, 1978, **80**, 281.
280. HUGH, R., and SAKAZAKI, R., *Publ. Hlth. Lab.*, 1972, **30**, 133.

281. VANDERZANT, C., and NICKELSON, R., *Appl. Microbiol.*, 1972, **23**, 26.
282. FISHBEIN, M., and WENTZ, B., *J. Milk Food Technol.*, 1973, **36**, 118.
283. COLWELL, R. R., LOVELACE, T. E., WAN, L., KANEKO, T., STACEY, T., CHEN, P. K., and TUBIASH, H., *J. Milk Food Technol.*, 1973, **36**, 202.
284. RICHARD, C., GIAMANCO, G., and POPOFF, M., *Annls. Biol. clin.*, 1974, **32**, 33.
285. BAUMANN, P., and BAUMANN, L., *J. Milk Food Technol.*, 1973, **36**, 214.
286. FISHBEIN, M., and WENTZ, B., *Microbiology—1974*, ed. Schlessinger, D., 1975, American Society for Microbiology, Washington, DC.
287. SAKAZAKI, R. *Food-Borne Infections and Intoxications*, 2nd Edn, eds. Riemann, H., and Bryan, F. L. 1979, Academic Press, London and New York.
288. SAKAZAKI, R., KARASHIMADA, T., YUDA, K., SAKAI, S., ASAKAWA, Y., YAMAZAKI, M., NAKANISHI, H., KOBAYASHI, K., NISHIO, T., OKAZAKI, H., DOKE, T., SHIMADA, T., and TAMURA, K., *Arch. Lebensmittelhyg.*, 1979, **30**, 103.
289. TAYLOR, J. A., MILLER, D. C., BARROW, G. I., CANN, D. C., and TAYLOR, L. Y. *Methods for the Isolation and Identification of Food Poisoning Organisms*, SAB Technical Series, in press.
290. SAKAZAKI, R., IWANAMI, S., and FUKUMI, H., *Jap. J. Med. Sci. Biol.*, 1968, **21**, 313.
291. WAGATSUMA, S., *Media Circle*, 1968, **13**, 159 (in Japanese).
292. WAGATSUMA, S., International Symposium on *Vibrio parahaemolyticus*, eds. Fujino, T. *et al.*, 1974, Saikon Publishing Co., Tokyo.
293. MIYAMOTO, Y., KATO, T., OBARA, Y., AKIYAMA, S., and TAKIZAWA, K., *J. Bact.*, 1969, **100**, 1147.
294. KANEKO, T., and COLWELL, R. R., *J. Bact.*, 1973, **113**, 24.
295. KANEKO, T., and COLWELL, R. R., *Microbial. Ecol.*, 1978, **4**, 135.
296. RUBY, E. G., and NEALSON, K. H., *Limnol. Oceanogr.*, 1978, **23**, 530.
297. DADISMAN, T. A., NELSON, R., MOLENDA, J. R., and GARBER, H. J., *J. Milk Food Technol.*, 1973, **36**, 111.
298. HOOPER, W. L., BARROW, G. I., and MCNAB, D. J. N., *Lancet*, 1974, **1**, 1100.
299. BEUCHAT, L. R., *J. appl. Bact.*, 1976, **40**, 53.
300. BEUCHAT, L. R., *Can. J. Microbiol.*, 1977, **23**, 630.
301. BEUCHAT, L. R., *J. Food Prot.*, 1977, **40**, 592.
302. RAY, B., HAWKINS, S. M., and HACKNEY, C. R., *Appl. Environ. Microbiol.*, 1978, **35**, 1121.
303. EMSWILLER, B. S., and PIERSON, M. D., *J. Food. Prot.*, 1977, **40**, 8.
304. MA-LIN, C. F. A., and BEUCHAT, L. R., *Appl. Environ. Microbiol.*, 1980, **39**, 179.
305. LEE, J. S., *J. Milk Food Technol.*, 1973, **36**, 405.
306. NELSON, K. J., and POTTER, N. N., *J. Fd Sci.*, 1974, **41**, 1413.
307. VANDERZANT, C., and NICKELSON, R., *J. Milk. Food Technol.*, 1973, **36**, 135.
308. ROLAND, F. P., *New Engl. J. Med.*, 1970, **282**, 1306.
309. ZIDE, W., DAVIS, J., and EHRENKRANZ, N. J., *Arch. int. Med.*, 1974, **113**, 479.
310. FERNANDEZ, R. C., and PANKEY, G. A., *J. Amer. med. Assoc.*, 1975, **233**, 1173.
311. HOLLIS, D. G., WEAVER, R. E., BAKER, C. N., and THORNSBERRY, C., *J. Clin. Microbiol.*, 1976, **3**, 425.
312. MCSWEENEY, R. J., and FORGON-SMITH, W. R., *Med. J. Austral.*, 1977, **1**, 896.
313. PROCIV, P., *Med. J. Austral.*, 1978, **2**, 296.

314. SAKAGUCHI, G., *Food-borne Infections and Intoxications*, 2nd Edn, eds. Riemann, H., and Bryan, F. L., 1979, Academic Press, London and New York.
315. HOBBS, G., *Adv. Fd Res.*, 1976, **22**, 135.
316. HUSS, H. H., *SIK Rapport*, 1976, No. 410, 3.
317. HUSS, H. H., *Appl. Environ. Microbiol.*, 1980, **39**, 764.
318. HOBBS, G., *Psychrotrophic microorganisms in spoilage and pathogenicity*, *Proc. XI IAMS Symposium*, eds. Roberts, T. A., Hobbs, G., Christian, J. H. B., and Skovgaard, N., Aalborg, Denmark, 1981.
319. HOBBS, G., JARVIS, B., and CROWTHER, J. S. *Methods for the Isolation and Identification of Food Poisoning Organisms*, SAB Technical Series, in press.
320. KROGH, G., *SIK Rapport*, 1976, No. 410, 73.
321. VAN ERMENGEN, E., *Z. hyg.*, 1897, **26**, 1.
322. SMELT, J. P. P. M., *Antonie van Leeuwenhoek*, 1973, **39**, 367.
323. HAUSCHILD, A. H. W., and HILSHEIMER, R., *J. Food Protect.*, 1979, **42**, 245.
324. SKOCZEK, A., *Medycyna wet.*, 1974, **30**, 48.
325. SKOCZEK, A., *Polskie Arch. Wet.*, 1976, **19**, 357.
326. ANDO, Y., OKA, S., and OISHI, K., *Bull. Jap. Soc, Sci. Fish.*, 1973, **39**, 505.
327. SASAJIMA, M., SHIBA, M., ARAI, K., and YOKOSEKI, M., *Bull. Jap. Soc. Sci. Fish.*, 1976, **42**, 469.
328. HUSS, H. H., SCHAEFFER, I., PEDERSON, A., and JEPSON, A., *Advances in Fish Science and Technology*, eds. Connell, J. J. *et al.*, 1980, Fishing News Books Ltd, Farnham, Surrey, England.
329. BANNAR, R., *Food Engineering International*, 1979, **4** (1), 56.
330. HUSS, H. H., SCHAEFFER, I., PETERSON, E. R., and CANN, D. C., *Nord Veterinaermd*, 1979, **31**, 81.
331. SKOCZEK, A., *Polskie Arch Wm wet*, 1976, **19**, 345.
332. DALLYN, H., and EVERTON, J. R., *J. appl. Bact.*, 1970, **33**, 603.
333. ZALESKI, S., CERONIK, E., SOBOLEWSKA-CERONIK, K., PENNO, J., and STARCZYK, L., *Acta Aliment. Pol.*, 1978, **4**, 177.
334. ARNOLD, S. H., and BROWN, D. W., *Adv. Fd Res.*, 1978, **24**, 113.
335. CRUIKSHANK, J. G., and WILLIAMS, H. R., *Br. Med. J.*, 1978, ii, 739.
336. GILBERT, R. J., HOBBS, G., MURRAY, C. K., CRUIKSHANK, J. G., and YOUNG, S. E., *Br. Med. J.*, 1980, **281**, 71.
337. HARDY, R., and SMITH, J. G. M., *J. Sci. Fd Agric.*, 1976, **27**, 295.
338. HAVELKA, B., *Prum. Potravin*, 1974, **25**, 24.
339. OMURA, Y., PRICE, R. J., and OLCOTT, H. S., *J. Food Sci.*, 1978, **43**, 1779.
340. LERKE, P. A., WERNER, S. B., TAYLOR, S. L., and GUTHERTZ, L. S., *West. J. Med.*, 1978, **129**, 381.
341. NIVEN, C. F., JEFFREY, M. B., and CORLETT, D. A., *Appl. Environ. Microbiol.*, 1981, **41**, 321.
342. GUTHERTZ, L. S., TAYLOR, S. L., LEATHERWOOD, M., and LIEBER, E. R., *Abstr. Ann. Meeting Amer. Soc. Microbiol.*, 1978, **78**, 189.
343. TAYLOR, S. L., GUTHERTZ, L. S., LEATHERWOOD, M., TILLMAN, F., and LIEBER, E. R., *J. Food Safety*, 1978, **1**, 173.

Chapter 4

THERMOBACTERIOLOGY OF UHT PROCESSED FOODS

K. L. BROWN and CELIA A. AYRES

*Campden Food Preservation Research Association,
Chipping Campden, UK*

SUMMARY

In this chapter an attempt has been made to summarise the literature relating to ultra high temperature (UHT) processing and the resistance of bacterial spores at high temperatures. Since the references on the subject are so numerous only a representative selection has been included. Throughout the chapter, to avoid confusion, temperatures are given in degrees Centigrade only, although much of the literature cited was written using the Fahrenheit scale.

DEFINITION

A UHT process has been variously defined as one above 121·1 °C with a maximum time of heating of 6 mins,[1] treatment at 135–150 °C for a few seconds,[2] processing with holding times of seconds to minutes at temperatures above 132 °C[3] or 126·7 °C,[4] or processing in the range 126·7 °C to 149 °C.[5]

While initially there may have been some distinction between UHT, HTST (high-temperature-short-time) and high-short[4] processing, these descriptions now appear to be interchangeable. Often, research into UHT processing has been linked to research into aseptic packaging, which is a method of food processing in which the product is sterilised prior to being sealed in a presterilised container.[6]

119

ADVANTAGES OF UHT PROCESSING

For many years the only way to pack and sterilise food was to put non-sterile food into non-sterile containers which were then sealed and heat processed[3] for sufficient time to destroy all the organisms and their spores capable of producing spoilage in the product. For those products in which the organism *Clostridium botulinum* could grow and form toxin, the process was designed to reduce the probability of survival of spores of this bacterium to less than 1 in 10^{12} containers containing the organism. In practice more severe processes were generally required to destroy the spores of more resistant spoilage organisms.

In conventional processing, the process time and temperature depends on the rate of heat transfer in the product and the container size. Increasing the container size increases the process required due to the longer time for the centre of the can to receive an adequate process. At high temperatures the product nearest the can wall receives an overcook while the product at the centre is just sterilised.[6] Products with a rapid rate of heat penetration, either due to convection currents within the can or by mechanical agitation of the can and contents, usually have a better quality if processed above 121·1 °C.[4] However, the maximum temperature for an agitated in-can process is around 132 °C due to the stresses on the container imposed by the pressures involved.[5] Some products are processed at this temperature, for example, cream style corn has been processed for 19 mins at 132 °C in a Sterilmatic rotary high-speed agitating cooker.[7,8]

To be able to process food at temperatures above 132 °C special equipment has been developed consisting basically of some form of heat exchanger to bring the temperature rapidly to 130–150 °C, a holding section to effect sterilisation and cooking, and a cooling device.[9] The product is then filled into containers aseptically.

The advantages of this system are numerous.[3,6] Off flavours caused by heat or container interaction are minimised. Colour retention is improved, vitamin retention is higher and the quality is constant irrespective of container size. Very large containers can be filled,[9] for example 55 gallon drums filled with banana pulp.[10] The types of container which can be employed are more varied than in the conventional process and include cans, glass jars, cartons, thermoformed plastic pots,[6] flexible plastic containers, steel drums and even rail cars.

Ball *et al.*[11] compared the quality of a range of products including soups, peas, beans, asparagus, evaporated milk and stews processed under HTST conditions with those processed conventionally. The HTST process was achieved by using thermal death time cans, 211 × 011 size, heated to 149 °C for 40–180 secs. The conventional process employed larger cans, 211 × 300 size, heated to 100–121·1 °C for 15–60 mins. They concluded that the flavour, colour, texture and vitamin content of the HTST foods were more like the unprocessed product than their conventionally processed counterparts. In a UHT plant, low acid products would be processed at 140–145 °C for 8–10 secs.[12]

Horner et al.[13] however, comparing the sensory aspects of UHT milk with whole pasteurised milk, found that over 75% of the tasters could accurately identify UHT milk with fewer liking UHT milk. UHT milk is normally processed to temperatures of 130–150 °C for 2–8 secs.[14] By raising the processing temperature and reducing the time, the bactericidal effect of the process is increased while at the same time the undesirable chemical changes, which cause browning, brought about by high temperatures, are reduced. There is approximately a threefold increase in the rate of browning in milk for each 10 °C rise in temperature while the same temperature rise

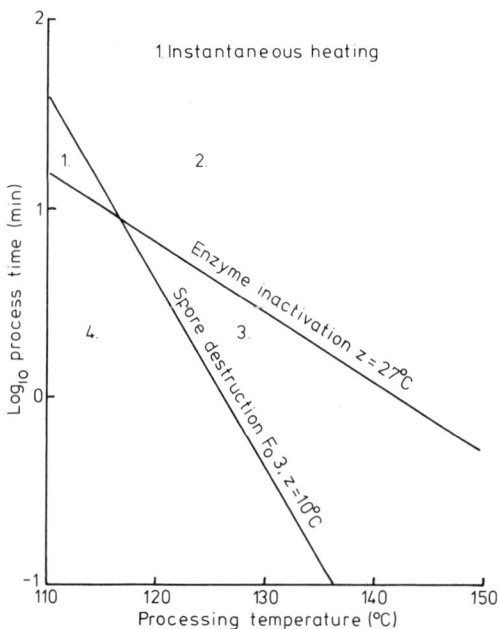

FIG. 1. Relative process times to achieve an $F_0 = 3$ for inactivation of bacterial spores and product enzymes in an instantaneously heating UHT system. (For explanation of numbers refer to text.)

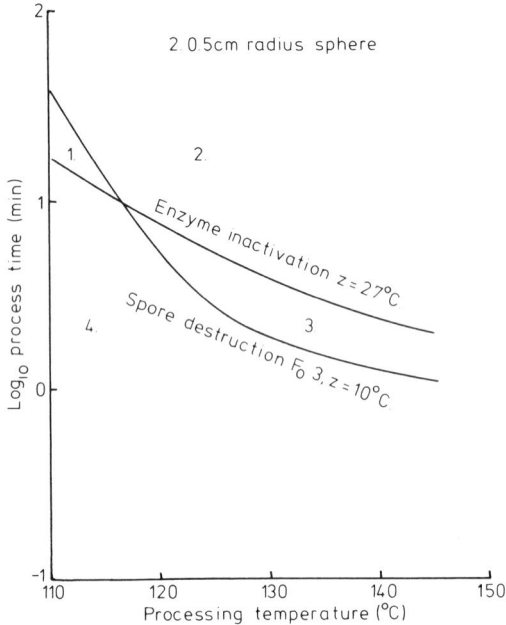

FIG. 2. Relative process times to achieve an $F_0 = 3$ for inactivation of bacterial spores and product enzymes in a food particle heated in a UHT system. (For explanation of numbers refer to text.)

increases the death rate of spores of *Bacillus stearothermophilus* and *Bacillus subtilis* by approximately 11 times and 30 times respectively.[2]

In any heat preservation treatment, it is necessary to inactivate the enzymes in the product (i.e. to cook it) as well as destroy bacterial spores. For an instantaneously heating system the rates of enzyme and spore inactivation can be represented by intersecting straight lines by plotting log process time against temperature. The slopes of the two lines are usually characterised by the term z value, which is the number of degrees Centigrade for the thermal destruction curve to traverse one log cycle.[15] This is illustrated in Fig. 1 for an F_0 3 process (F_0 is the equivalent in minutes at $121\cdot1\,°C$, when $z = 10\,°C$, of all heat considered, with respect to its capacity to destroy spores or vegetative cells of a particular organism).[15] The process time/temperature combinations fall into one of four sectors:

(1)   product not sterile, enzymes inactivated;
(2)   product sterile, enzymes inactivated;
(3)   product sterile, enzymes not inactivated;
(4)   product not sterile, enzymes not inactivated.

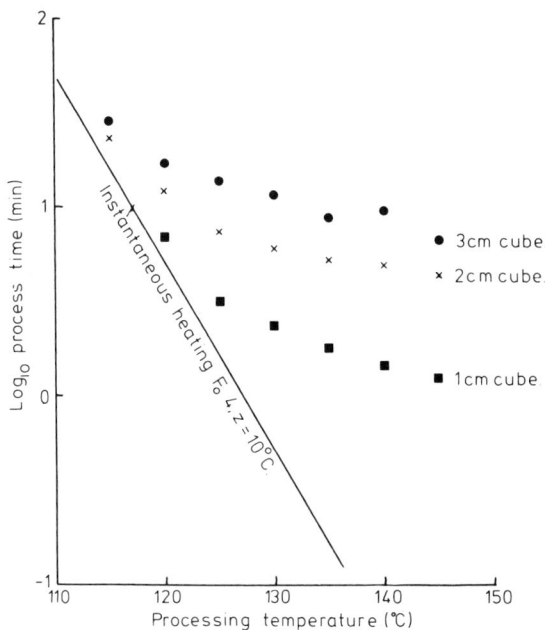

FIG. 3. Experimental process times required to achieve $F_0 = 4$ in potato cubes of different sizes heated in steam between $110\,^{\circ}C$ and $150\,^{\circ}C$.[17]

Therefore at high temperatures the product may be sterilised but remain uncooked[9] if the enzymes are not inactivated.

When solid particles are subjected to UHT processing, the holding time has to be extended to ensure that the particles receive an adequate process because of the time taken for heat transfer into the particle. In a trial sterilisation using 25 mm meat cubes[16] it was found that a minimum holding time of 5 mins at $135\,^{\circ}C$ was required compared with 2 secs at $135\,^{\circ}C$ for a UHT milk product.

If log process time is plotted against temperature for a $0\cdot5$ cm radius sphere (Fig. 2) the enzyme and spore inactivation rates become curves because of the time required for the heat to penetrate the sphere. Again the four time/temperature sections apply as in Fig. 1.

Newman and Steele[17] demonstrated the effect of particle size by determining experimentally the times to achieve F_0 of 4 in different size potato cubes. These times are plotted in Fig. 3. The larger the cube size the longer the process time required to achieve an equivalent process in the UHT range. For a 3 cm cube heated at $135\,^{\circ}C$ this would mean a process time around 100 times as long as that required by an instantaneously heating fluid.

COMMERCIAL SYSTEMS

Commercial UHT systems fall into two categories, indirect or direct heating. In indirect systems the product does not come into contact with the heating medium while direct systems achieve heating by pumping product into steam or injecting steam into product.

Indirect Heating

In an indirect system because of the exponential rate of temperature rise probably 30–60 % (depending on the length of the holding time) of the total sporicidal and chemical effects take place during the relatively long heating-up period at relatively low temperatures. Temperatures in the range 140–150 °C cannot normally be obtained without overheating.[2] Indirect systems suffer from local hot spots and burning on of products,[18] and are therefore usually used for thin products. The heat exchangers are of the tubular, plate or swept surface type.

Tubular heaters, for example the Stork, Mallorizer, Roswell, Sterideal and Spiratherm types,[2,9] consist of heated tubes along which the product is pumped.

Plate type heat exchangers, for example the APV Ultramatic, Alfa-Laval VTS, Paasch and Silkeborg and Rosenblad types,[2,9,12,16,19,20] consist of closely spaced plates with corrugations to produce turbulence. Some designs work up to 150 °C when synthetic rubber gaskets are used between the plates.[12] Heat transfer coefficients are greater for a plate type than a tube and residence times are usually shorter.[16] The APV Ultramatic has been used for processing milk at a temperature of 138 °C for 1·5 secs with cooling to below 100 °C achieved in less than 1 sec.[16]

Scraped or swept surface heat exchangers, for example the Johnson Votator, the APV-Clark Rota-Pro and the Cherry Burrell Thermutator, have been used for more viscous liquids such as egg concentrate, tomato paste and sauces.[12] Cream style corn has been processed in a Votator at a temperature of 143 °C with a 30 sec hold time to achieve an F_0 of 30,[21] while a more unusual product to be successfully processed in this way is banana pulp.[10,22]

Direct Heating

In a direct heating system the product is heated by mixing it with steam under pressure. Heating and cooling is virtually instantaneous and almost all of the sporicidal and chemical effects take place at full processing temperature. This method also allows the use of higher temperatures.[2] Direct heating methods are generally grouped into two types; steam into product and product into steam.

Steam injection into product systems, for example the Uperisation, Alfa-Laval VTIS and Cherry Burrell plants[2,9,16,18,19] are generally used for low viscosity products since turbulence aids heat transfer.[9] This method of heating is often used for milk. Steam injection in the Uperiser raises the temperature of milk from 80 to 150 °C with a nominal holding time of 2·4 secs.[16] If the water content of milk is to remain the same, then the amount of water added as condensate during steam heating must equal the amount removed during expansion cooling.[23]

Steam injection systems operating at 135–163 °C with holding times of 3–13 mins have been used to sterilise fermentation media.[24,25]

Product into steam apparatus is more flexible than steam injection systems since the flow rate is more stable and there is less danger of overheating.[9] Examples of this type of heat exchanger are the Laguilharre, Thermovac, Palarisator, Ultratherm and Calefactor.[2,9]

Product into steam equipment has been used to process large particles. The Abrams process used for meat slices, the Smith–Ball process operating at 138–149 °C for chop suey and the Martin process operating also at 138–149 °C are three examples of this.[26] An interesting point is that the Martin process took 8–10 secs at 141 °C to process pea soup but took 38–60 secs to process c. 1 cm cubed vegetables which again demonstrates the effect of particle size on the process required.

IMPORTANCE OF RESIDENCE TIME

In a continuous UHT steriliser some product travels relatively slowly through the system and is subjected to a longer heating time while other product travels more rapidly. The spread of velocities depends on the degree of turbulence.[27] The residence time distribution has a very important effect on the number of survivors of a heat treatment and the F_0 value of the process.[28–31]

Bateson,[29] using a salt solution impulse test, examined the effect of rotation speed, flow rate and fluid viscosity on the holding time distribution for a swept surface heat exchanger. Other methods of determining residence times that have been reported include use of dyes[32] and fluorocarbon tracers.[33] Neglect of velocity profiles in continuous sterilisers could increase the probability of survival of sporeformers.[34]

PROCESS DETERMINATION

In the canning industry, processes are designed to ensure a reduction in the probable survival of *Cl. botulinum* spores in low acid foods by a factor of at least 10^{12}. Speck and Busta[35] calculated that a process with a sterilising effect of 12 log reductions for *Cl. sporogenes* at 148·9 °C would be 3·42 secs. This was from a spore suspension characterised by a z value of 19·4 °C and a decimal reduction or D value at 148·9 °C of 0·285 secs. It was calculated that this process would be well above the requirements for destruction of *Cl. botulinum* spores described in the literature.

In the dairy industry a spoilage rate of a maximum of one container in a thousand appears to be acceptable[36] for UHT processing. In practice, better results than this are usually achieved. Raw milk normally contains approximately one resistant spore per ml[37] and a UHT process can easily produce a sterilising effect of 7 to 8 log reductions in number of spores. It can often be assumed therefore that spoilage, when it occurs, arises by contamination during the filling and closing operation after the process or by container contamination. When larger volume containers are employed it may be necessary to increase process levels to achieve the same spoilage rate of less than one container in a thousand.

Ashton[38] described the setting up of a UHT-Tetra Pak system for UHT milk. Initially, proving trials using water were carried out to check that there were no leaks and that the timing sequences were working properly. This was followed by sterilisation trials at 143–149 °C using skim milk. If this proved satisfactory then the process temperature was determined by dropping the temperature in 2·5 °C intervals and checking samples for sterility. The 'safe' temperature was set 2·5 °C above the lowest temperature which gave no survival based on an average of three trials.

Occasionally, spoilage may arise due to an unexpected reason. Luck *et al.*[39] reported spoilage in samples of UHT milk processed at 145–150 °C by the direct method in a commercial plant in South Africa. After an incubation period of 5–7 days at 35 °C, 72% of the defectives tested contained pure cultures of mesophilic sporeformers. Process survival was attributed to the fact that the raw milk in South Africa contained up to

13 600 spores/ml whereas European milk contains usually less than 1 spore/ml.

KINETICS OF DEATH OF BACTERIAL SPORES

It is perhaps appropriate at this point to summarise briefly the types of spore survivor curve that may be obtained and discuss the methods of presentation of spore heat resistance data. These aspects have been reviewed in more detail elsewhere.[1,15,40-42]

Traditionally the decrease in the number of viable spores is considered to be an exponential function of time at a constant temperature so that plotting log survivors against time gives a straight line.[5] In practice, however, a straight line is rarely obtained.[15,41-43] Figure 4 illustrates some of the survivor curves that have been obtained. Williams et al.[44] and Pflug and Smith[45] obtained concave downwards survivor curves for B. stearothermophilus while Shull and Ernst[46] reported an initial activation shoulder for the same species heated in saturated steam on paper strips. Pflug and Smith[45] found that the shape of the survivor curves for Cl. sporogenes varied depending on the heating medium and the test temperature between 105 and 121 °C. At low temperatures the survivor curves were sigmoidal, at middle temperatures they were straight and at the

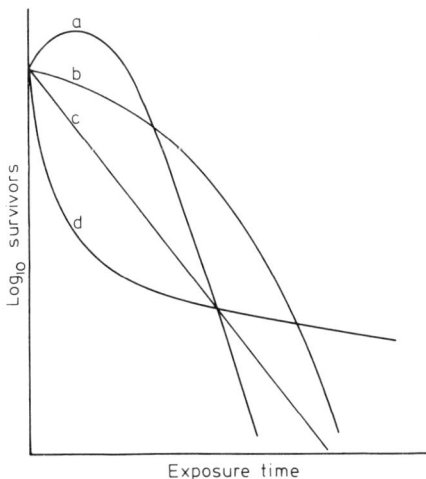

FIG. 4. Typical survivor curves for bacterial spores exposed to moist heating conditions. (a) Initial activation or deflocculation shoulder; (b) concave downwards; (c) logarithmic order of death; and (d) tailing curve.

highest temperatures they were concave downward. The composition of the sporulation medium has been shown[47] to affect the shape of survivor curves of *B. stearothermophilus*.

Tails on survivor curves are often reported. In capillary tubes, Bean *et al.*[48] attributed tailing to an artifact of the method. Tailing may also be due to variation in heat resistance within a population of spores but this is more likely due to physiological rather than genetical differences.[49]

Many attempts have been made to explain non-logarithmic survivor curves. Moats[50] argued that death could not follow first order kinetics since this implied inactivation of a single site per cell which failed to explain sublethal injury. He therefore assumed that thermal death resulted from the inactivation of a proportion of a number of critical sites within the spore. Humphrey and Nickerson[51] concluded that survivor curves were non-logarithmic and postulated that death was due to damage to a heat-induced, heat-labile reproduction initiator. The distribution of resistance theory was discounted by Charm[52] since it would involve a skewed distribution, while Aiba and Toda[53] concluded that a logarithmic death rate was the most probable representation of the individual life span distribution of spores.

Faced with these problems, some researchers have tried to describe survivor curves by a single straight line regardless of shape. Others[45] have used formulae which assume a straight line for example

$$\log N_U = - U/D + \log N_0$$

where N_U = number of survivors after process of time U, $D = D$ value, N_0 = initial number of spores. In this case, if the survivor curve had an initial shoulder, the calculated D value would be lower than that obtained by taking the reciprocal of the slope of the straight line portion of the survivor curve.[5] Conversely, using the D value obtained from the straight line portion of the curve and using this value to work backwards to find N_U would lead to an erroneously high inactivation factor.[54]

Shull *et al.*[55] attempted to describe a survivor curve with an activation shoulder by the equation

$$A_t = A_0 \exp\left(-kt\right) + \frac{\alpha N_0}{k - \alpha}\left(\exp\left(-\alpha t\right) - \exp\left(-kt\right)\right)$$

where A_t = number of spores activated at time t, A_0 = number activated at $t = 0$, α = activation rate constant, k = death rate constant and N_0 = number of non-activated spores. Others[56] have straightened concave downwards curves by allowing for heating come-up time.

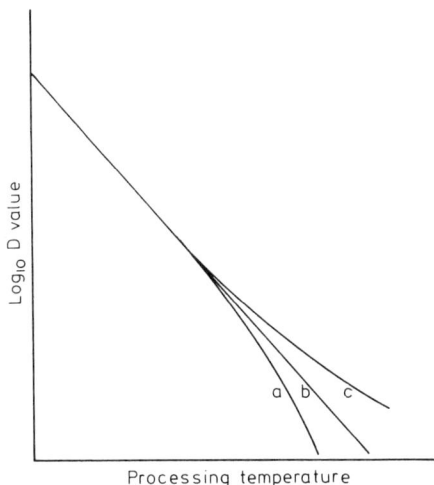

FIG. 5. Thermal death time curves reported for bacterial spores exposed to moist heating conditions. The z value (a) decreasing; (b) linear; and (c) increasing with temperature.

Again, traditionally, when log D value is plotted against temperature a straight line, designated the thermal destruction curve, is obtained, from which the z value is derived.[5] However, Gillespy[57] plotting log D values for *Clostridium thermosaccharolyticum* against temperature and the reciprocal of temperature concluded that the z value varied continuously with temperature. He analysed the data of Esty and Meyer[58] for *Cl. botulinum* and found that a constant activation energy gave predicted D values closer to the experimental than a constant z value.

Wang *et al.*[59] correlated the death rate data for *B. stearothermophilus* obtained over the temperature range 127·2–143·8 °C by both the Arrhenius and the Thermal Death Time theory, the equations for which are

Arrhenius: $k = A_e^{-E/RT}$

Thermal Death Time: $\log (D/D_{121·1}) = -(t - 121·1)/z$

where k = specific reaction rate, A_e = Arrhenius constant, E = activation energy, R = gas content, T = absolute temperature, $D = D$ value, $D_{121·1} = D$ value at 121·1 °C, t = temperature in °C and z = thermal death time constant. A straight line relationship was obtained with an Arrhenius plot but the thermal death time plot produced a concave upwards curve. Other workers[48,60,61] have obtained concave upwards thermal death time curves for *B. stearothermophilus*. However, for *Cl. sporogenes*, both concave upwards[45,62] and downwards[45,63] thermal death time curves have been reported. These types of curve are illustrated in Fig. 5. Applying the Arrhenius equation to concave downwards thermal death time data would increase the curvature.[40]

Over a limited temperature range both methods may be used[63-65] but outside this range the Arrhenius approach will predict longer D values.

FACTORS AFFECTING SPORE RESISTANCE

The factors affecting spore heat resistance data have been adequately reviewed elsewhere[40-42] and therefore only a few examples will be given here.

Factors affecting the survivor curves of spores tend to fall into four main categories namely:

(1) those affecting sporulation;
(2) those occurring during storage of the spore suspension;
(3) those affecting the conditions during the heat treatment;
(4) those affecting the recovery of survivors.

The methods employed to produce spores vary according to species and the composition of the sporulation medium is often a major contributive factor to the degree of resistance, the shape of survivor curve, and the z value.[47,66]

Various sporulation medium additives have been reported to increase spore resistance. For example, manganous ions have been shown to increase the resistance of spores of *B. stearothermophilus*[67] and *Bacillus coagulans* var. *thermoacidurans*[68] and long chain fatty acids to increase the thermal resistance of *Cl. botulinum*.[69]

Sporulation temperature can have a dramatic effect on resistance. Cook and Gilbert[67] observed a doubling of the D value at 115 °C from 12·2 to 24·4 mins for *B. stearothermophilus* when the sporulation temperature was increased from 50 °C to 60 °C.

Since large volumes of spore suspension are usually prepared in advance of a series of resistance studies, the effect of storage becomes important. Davies *et al.*[60] stored their spore crops of *B. stearothermophilus* for at least 6 months at 4 °C because less consistent results were obtained with freshly harvested spores. It has been shown[70,71] that prolonged storage of *B. stearothermophilus* spores reduced the activation required before germination occurred and lowered the resistance.

During the actual heating of the spores, one of the major factors affecting the resistance is the pH of the suspending medium. Maximum resistance is usually around the point of neutrality[67,72-74] at normal processing temperatures but under UHT conditions, Brown and Thorpe[73] predicted that this may not be the case.

In other experiments the addition of sugars to the suspending fluid has been shown to increase the resistance of spores of *Cl. sporogenes*[75] and *Cl. botulinum.*[69]

Following heat treatment, surviving spores have usually suffered some degree of heat damage. The type of recovery media employed[76-78] together with the incubation temperature[79] has an important bearing on the percentage of heat damaged spores enumerated.

METHODS OF DETERMINING BACTERIAL SPORE HEAT RESISTANCE

Before the processing methods described earlier can be considered safe for commercial production, determinations of the sterilising effects are made. Initially this involves physical measurements of the temperature profiles, and flow rates where applicable. The data obtained from these measurements are usually consolidated with spore resistance studies both in the laboratory and on the actual commercial equipment. It is not recommended that UHT process values are determined by extrapolating spore heat resistance data obtained at traditional processing temperatures (110–125 °C) to the higher temperature range. Process calculations frequently assume a linear z value for spores whereas non-linear z values are often reported.[45,59,62,63]

At very high temperatures the death rate of bacterial spores is so rapid that special techniques have been developed to study their resistance. Very often a number of assumptions regarding heat transfer and corrections for heating and cooling lags have to be applied to the data. The corrections applied are often quite large in relation to the death rate obtained. Ideally, heating and cooling times should be negligible compared with the holding time.[80]

The types of equipment used for determination of spore resistance at high temperatures are many and varied. They have been grouped here under four headings: indirect heating methods; mixing methods; direct heating methods and particle methods.

Indirect Heating Methods

In indirect heating methods temperatures generally rise exponentially so that a significant proportion of the heating-up or come-up time is at the higher temperatures where lethality is greatest. This is probably the main limiting factor of this type of system since the spores may well have been killed before the operating temperature has been reached. (In theory operating temperature is never actually attained.)

Thermal Death Time and Capillary Tubes

Probably the simplest method of determining spore resistance was the thermal death time (TDT) tube technique of Bigelow and Esty[81] developed in 1920. Spore suspension was sealed into tubes of 7 mm i.d. and 1 mm wall thickness which were immersed in an oil bath held at constant temperature. Single tubes were removed at defined time intervals and cultured to show presence or absence of growth.

Esty and Williams[82] developed a multiple tube method so that percentage survival could be calculated. Straight lines were obtained when log percentage survival was plotted against time.

By using smaller diameter capillary tubes which could be accurately filled using a microsyringe[83] later experimenters were able to determine spore resistance at higher temperatures.

Franklin *et al.*[84] using 75 mm × 0·8 mm i.d. capillary tubes working with *B. subtilis* in milk or water heated between 110 °C and 120 °C obtained Q_{10} values ranging from 16 to 35 depending on the strain. The z value is related to Q_{10} by the equation:[1]

$$z(°C) = \frac{10}{\log Q_{10}}$$

The survivor curves tailed off at low survivor levels and the thermal death curves were sigmoid rather than exponential. The results were, however, used to construct a theoretical destruction curve, by the method of Burton,[27,28] to predict the survival of spores heated to 135 °C in an APV HX indirect heat exchanger. The predictions agreed reasonably well with actual results from the plant.[85]   Further experiments were performed with *B. stearothermophilus* TH24 heated in capillary tubes between 110 °C and 125 °C,[86] in which logarithmic survivor curves were obtained. The Q_{10} value for spores of *B. stearothermophilus* in water was *c*. 20 and in milk *c*. 11·5 which differed from the earlier results[84] with *B. subtilis* where the Q_{10} in water was *c*. 16 and in milk *c*. 30.

Temperatures inside a capillary tube can be calculated[87,88] assuming it to be an infinite cylinder using the procedure of Olson and Schultz.[89] The lethality inside the tube is calculated by assuming a number of imaginary concentric layers within the tube. Each layer is bounded by an iso-j region where j is the heating or cooling lag factor.[15] Using a small copper–constantan thermocouple, 0·34 mm o.d., connected to an ultraviolet recording oscillograph, Perkin et al.[56] were able to measure temperatures within a capillary tube. Correcting survivor curves of B. stearothermophilus spores[60] for come-up time had the effect of straightening the thermal death time and Arrhenius plots. It was concluded that with heating times greater than 5 secs, the heating-up period would have negligible effect on the determination of the slope of a thermal death time curve. Above 135 °C heating times of less than 5 secs would be involved and corrections necessary. It should be noted that in the authors' experience there are a number of different sizes of capillary tube available. Since heating lag calculations involve a radius term,[87–89] size variations are a factor to be considered.

Spores, of B. stearothermophilus (NCDO 1096) heated in capillary tubes[60] between 120 °C and 160 °C had Q_{10} values of 23·5 in water and 13·2 in milk. Survivor curves showed an increase in sigmoid shape as heating temperature increased, and the lack of a heat activation peak for spores of this organism as observed by other workers[46] was noticeable. Thermal death time curves were also concave upwards at high temperatures. After computer correction[56] for come-up time the degree of curvature was reduced but the curve was not entirely straightened. Neaves and Jarvis[90,91] applied the same computer correction program to their data for spores of Clostridium botulinum type A heated in capillary tubes between 85 °C and 160 °C. Even after correction there was a pronounced upward curve to the thermal death time plot indicating a significant increase in z value with an increase in temperature. At this time this is the only piece of UHT data available for spores of Cl. botulinum and clearly further information is needed if the safety of UHT processes are to be ensured.

In order to simplify the cooling lag corrections for capillary tubes, Stern and Proctor[92] in 1954 devised an apparatus consisting of weights, levers and pulleys to transfer the tubes rapidly from the oil bath to the cooling tank. Stern et al.[93] had earlier used a copper–constantan thermocouple connected to a cathode ray oscilloscope to monitor temperature changes in capillary tubes. Using the methods described by Olson and Jackson[94] it was calculated that the centre of a tube would take 8·31 secs to reach 131·4 °C in an oil bath at 135 °C.[92]

Later modifications to the apparatus utilising pneumatic jacks[95] and spring operated levers[96,97] have been reported. Using the spring operated

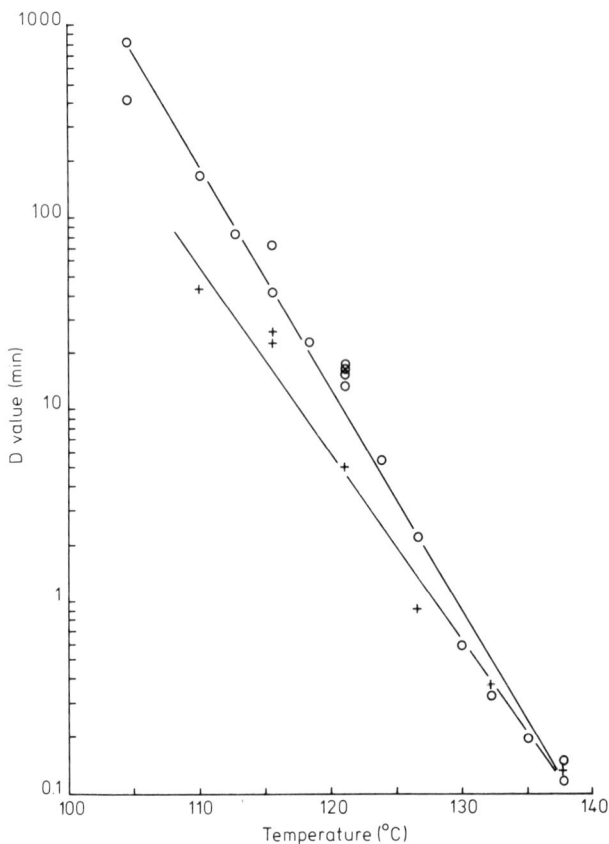

FIG. 6. Thermal death time curves f r spores of *Bacillus stearothermophilus* heated at pH 6 (○) and pH 7 (+), using capillary tubes in the Stern and Proctor system.[97]

system, linear thermal death time curves were obtained for *B. stearothermophilus* spores heated between 104·4 °C and 137·8 °C (Fig. 6) at pH 6·0 and 7·0.[97] The *z* values obtained were 8·7 °C and 10·3 °C respectively.

An interesting variation in heating technique was reported by Reichart.[64,98] The traditional approach has been to heat tubes in a constant temperature oil bath at a number of temperatures to establish a thermal death curve from which *z* or Q_{10} values could be calculated. By using an exponentially heating system with tubes removed at successive time intervals Reichart was able to construct a survival curve for *B.*

stearothermophilus spores with the slope increasing with time as the temperature increased. Death rate coefficients were calculated from the slopes of the curve at a number of temperatures in order to construct a thermal death time or Arrhenius curve from which z or activation energy could be obtained. This method provided values for decimal reduction time (D) and z in one experiment rather than several.

The curve for *B. stearothermophilus*[64] given in the paper showed no evidence of activation of the spores by heat. The authors have tried this approach (unpublished data) heating *B. stearothermophilus* spores in capillary tubes in an oil bath heated electrically to obtain a linear temperature rise.[25] Both an activation and a destruction curve were obtained.

Two methods based on the capillary tube approach have been used to assess processes in cans processed in continuous sterilisers, while neither method has been used on the UHT range, they are included here for interest and completeness.

Spores of *B. stearothermophilus* sealed in capillary tubes or capillary bulbs[99-101] which are then placed inside cans of food have been used to evaluate the process received at the centre of each can (F_c value).[15] This approach has been used in continuous cookers, where conventional thermocouple systems cannot be used, up to the normal maximum can processing temperature of around 130 °C.

Pflug *et al.*[102] used *B. stearothermophilus* spores sealed in plastic rods to assess processes in an FMC Steritort. The rods were positioned in cans using Ecklund receptacles. The F_0 values calculated from spore survival were in close agreement with those from thermocouple measurements. It was suggested that if five cans were set up in this way, the average spore F value would be within 7% of the true value 95% of the time.

Thermal Death Time Cans
Another widely used indirect heating technique was the thermal death time (TDT) can method developed by Townsend *et al.* in 1938.[103] Spores were heated in 208 × 006 (2·5 inch × 0·375 inch) cans containing approximately 13 ml of food product. The cans were heated in the temperature range 100–142 °C in specially constructed miniature retorts[103-105] which had a rapid come-up time. Heat resistance studies were often carried out in parallel in pyrex thermal death time tubes.[103,105] Schmidt[106] developed a method of heating cotton plugged tubes containing 0·1 ml spore suspension and 1 ml of product in a miniature retort constructed from a 6 inch diameter pipe supplied with steam from a 30 gallon ballast tank. Come-up times for

the tubes were of the order 0·6 to 1·1 mins depending on product, with a maximum temperature of 131 °C for the system.

In their apparatus, Townsend *et al.*[103] allowed a 0·33 min come-up time correction for cans based on the work of Richardson[107] and 0·73 to 0·93 mins for tubes from the calculations of Ball *et al.*[108] Other workers[105,109] have applied corrections ranging from 0·85 to 1·7 mins for tubes and 0·55 to 1·93 mins for cans depending on product. Sognefest and Benjamin[109] found that with a convection product the orientation of the can in the retort affected the rate of heat penetration.

Spore resistance data obtained using the thermal death time can method has generally been limited to temperatures below the UHT range partly because of the come-up time although some experiments up to 142 °C have been reported.[104]

Linear thermal death time curves have been obtained for *Clostridium sporogenes* 3679 and *Cl. botulinum* over the temperature range 100–126·7 °C. The z values reported were of the order 5·6–12·8 °C for *Cl. botulinum* 62A spores and 4·4–11·1 °C for *Cl. botulinum* type B spores depending on the product in which they were heated.[103–105] Decimal reduction times at any one temperature also varied depending on heating substrate.

Continuous Flow Indirect Heating Systems
Both tubular[110–113] and plate type heat exchangers[85,86,114] through which spore and substrate are pumped continuously have been employed to determine spore resistance at high temperatures. The apparatus used, in some cases, is very similar to commercial plant.[85,86,111–114] The main advantages of using a continuous flow system over the tube or can methods just described are that a large volume of spore suspension can be processed at high temperatures with rapid heating and cooling, and the effect of residence time within the system on spore survival can be studied. The importance of residence time has been discussed earlier.

Martin[113] inoculated various food products with spores of putrefactive anaerobe 3679 and heated them in a simple coiled tube arrangement at temperatures between 130 °C and 148 °C. The product was then canned in a Martin aseptic canning unit, and incubated to determine spoilage levels. After processes equivalent to F_0 6·2–6·7, spoilage levels of 98 %, 75 % and 4·2 % were recorded in puréed peas, bacon soup and liver soup respectively. Some of this variation was undoubtedly due to differences in heat transfer and flow characteristics.

Galesloot[111] expressed the sporicidal effect of the process given by a

Stork tubular heater in terms of a 'sterilising effect' given by

$$\text{sterilising effect} = \log\left(\frac{\text{initial concentration}}{\text{final concentration}}\right)$$

A sterilising effect ranging from 0 to 6 for *B. subtilis* was obtained in the Stork heater when the operating temperature ranged from 120 °C to 135 °C. Increasing the holding time by 16·5 secs increased the sterilising effect by approximately 2.

A stainless steel tubular heater has also been used to determine the conditions for inactivation of milk borne foot-and-mouth disease virus at ultra high temperatures.[110] With temperatures up to 148 °C and holding times of 2–3 secs it was found that a temperature of 148 °C was required to destroy the virus. The heat resistance of viruses in the UHT range appears to have largely been ignored and in view of the resistance reported here, perhaps this is an area for future research.

A continuous flow tubular system employing slug flow, achieved by mixing air bubbles with the product as in an auto analyser, has been reported[115] for use in the temperature range 50–72 °C. Holding times were of the order 2 to 60 secs with heating and cooling periods of 1·7 secs. It is possible that such a system could be suitably modified for use at high temperatures.

A 900 litre per hour APV HX plate type heat exchanger was evaluated by Williams *et al.*[114] for sporicidal effect using *B. subtilis* 786 spores in water. A sterilisation effect of 7 was obtained with an operating temperature of 135 °C and a holding time of 2–4 secs. When the work was repeated using milk, a temperature of 130·5 °C was found to give the same sterilising effect.[85] This apparent increase in efficiency was found to be due to substances present in the milk inhibiting spore germination.

To achieve a sterilising effect of 7 for spores of *B. stearothermophilus* TH24 in milk heated in the same APV heat exchanger required a temperature of 142 °C with a holding time of 2 secs.[86] At a temperature of 138 °C the sterilising effect after 2 secs was only 2. When the inhibitory effect of milk on germination was not taken into account the sterilising effect appeared to be 8.

Mixing Methods

One approach to reduce the time taken for a spore suspension to reach processing temperature is to mix a very small volume of liquid containing spores with a much larger volume of substrate which has been preheated to the required temperature. Provided that mixing takes place rapidly, the spores are brought to temperature almost instantaneously. One drawback to this approach is that it is limited to non-viscous liquids.

Aiba and Toda[53] used this method to obtain rapid heating by injecting 1–5 ml of spore suspension into 15 litres of preheated buffer solution while Kooiman[116] injected 0·1 ml of sample into 10 mls of previously equilibrated heating medium in screw-cap tubes. The temperatures employed in both these systems were below 100 °C but with suitable modifications higher temperatures could be achieved.[117]

This procedure has also been adopted for continuous flow[59,80] by mixing water, preheated to high temperatures, with spore suspension in appropriate volumes in a pressurised mixing chamber. The flow was induced by means of compressed gases and flow rates were determined using calibrated rotameters. Mixing times as short as 0·6 msecs were reported,[59] while exposure time could be controlled by altering the flow rate or reactor length. The spore suspension was flash cooled by passing it into an expansion chamber. Residence time distributions were determined using salt solution by observing concentration changes at the outlet due to a forced step change in inlet salt concentration.

Death rate data were obtained for spores of B. stearothermophilus over the temperature range 127·2 °C to 143·8 °C by Wang et al.[59] in the continuous flow system and were compared with data from lower temperatures using the capillary tube technique. Activation energies and Arrhenius constants were similar for both sets of data but extrapolation of the capillary tube data to higher temperatures using a thermal death time curve showed nearly 160% disagreement with the reactor data. It was concluded that a thermal death time curve was a poor way to correlate data over an extended temperature range.

Oquendo et al.[80] obtained a straight line relationship by plotting log D against temperature for B. stearothermophilus spores heated in their apparatus between 120 °C and 150 °C. The z value was 5·6 °C with a D value at 120 °C of approximately 10 secs which is low for this organism. One drawback of this particular apparatus was the long residence time.

It appeared to be possible using either continuous flow system to measure D values as small as 0·001 to 0·006 of a sec even though the shortest holding time in one apparatus[59] was 0·2 sec while the mean holding time in the other[80] ranged from 15 to 60 secs.

Direct Heating Methods

These techniques take advantage of the latent heat of steam to produce extremely rapid heating. Experiments have been carried out in both product into steam and steam into product systems, some in commercial scale equipment.

Product into Steam Methods

The thermoresistometer. The thermoresistometer was an apparatus designed by Stumbo[118] in 1948 specifically to determine the thermal resistance of bacterial spores in the temperature range 101·6 °C to 132·2 °C. A similar apparatus was produced by Pflug and Esselen[119,120] to operate up to 148·9 °C, and a later design[121] was developed for both wet and dry heat resistance studies.

The basic procedure consists of heating 0·01 to 0·02 ml of inoculum plus substrate in five or six small metal cups in a specially constructed steam chamber. Following heating, the cups are subcultured in suitable media. It is a rather specialised piece of equipment and as such is rather expensive to construct, but, once built, large numbers of replicate samples can be processed in a relatively short time. Substrates are generally confined to liquids and food homogenates. The relative advantages and disadvantages of this method are listed by Stumbo[15] and Pflug and Schmidt.[40]

The thermoresistometer was used initially for thermal resistance studies on spores of PA (Putrefactive Anaerobe) 3679[118] suspended in various food products in the temperature range 104·4 °C to 132·2 °C. This work was later extended[122] to include *Cl. botulinum*. The semi-logarithmic thermal death time curves obtained were straight lines.

Pflug and Esselen[119,123] also working with PA 3679 but heated in M/15 neutral phosphate buffer, obtained a linear relationship between log D value and temperature between 112·8 °C and 148·9 °C, with a z value of 9·3 °C and D value at 121·1 °C of 1·06 mins. However, later experiments[62] in which PA 3679 spores were heated in a variety of substrates suggested that the z value increased with temperature.

Secrist and Stumbo[124] disagreed with this and set out to determine the resistance of PA 3679 spores in strained peas and distilled water at temperatures of 121 °C to 143 °C. Both before and after corrections for heating and cooling lags had been applied, the thermal death time curves were virtually linear. It was concluded that, with heating times as short as 0·021 mins at 143 °C, it was scarcely short of fortuitous to obtain an exact value for thermal resistance. The original apparatus was, after all, only intended for use up to 132·2 °C.

Attention was drawn by Secrist and Stumbo[125] to variations in D values obtained for PA 3679 spores heated in various products and the implications for process calculations.

TABLE 1

SUMMARY OF UHT THERMAL RESISTANCE DATA OBTAINED USING THE THERMORESISTOMETER

Organism	Suspending medium	$D_{121\cdot1^{\circ}C}$ (min)	z (°C)	Temperature range (°C)	Reference
Cl. sporogenes PA 3679	Pea purée	1·68	9·44 (linear)	104·4–132·2	118
	Several substrates	0·8–1·52	9·2–11·4 (linear)	104·4–132·2	122
	Phosphate buffer (pH 7)	1·06	9·3 (linear)	112·8–148·9	119
	Several substrates	0·75–2·03[a]	9·0–14·7[a] (non-linear)	121·1–143·3	62
	Water	0·63–0·73	10·4 (linear)	121·1–143·3	124
	Strained peas	0·95–1·25	9·8 (linear)	121·1–143·3	124
	Several substrates	0·24–0·58	9·4–10·4 (probably linear)	110·0–132·2	125
Cl. botulinum	Several substrates	0·051–0·133	8·2–9·1 (linear)	104·4–126·7	122

[a] Stumbo's method.[80]

The results of the experiments just described are summarised in Table 1. The thermoresistometer has also been used to study the thermal destruction of spores of *Bacillus coagulans* at lower temperatures.[126,127]

Laboratory scale product into steam UHT plant. Perkin[128] reported the development of a laboratory scale batch steriliser which enabled extensive spore heat resistance studies to be carried out in water and in milk, whilst being relatively economical on spore stocks. The steriliser was designed to process small volumes of sample (up to 500 ml) at 130–150 °C for a few seconds, with negligible come-up time. This equipment is often referred to in the United Kingdom as the MISTRESS (milk into steam treatment research equipment small scale).

The steriliser consisted essentially of a heating chamber and a flash-cooling chamber.[101] The sample of milk or water containing spores was introduced into the steam in the heating chamber as a fine spray. The holding time was controlled by varying the time between the injection of the sample and the opening of the valve into the cooling chamber. Cooling was achieved mainly by evaporative cooling in the second chamber.

The sterilising effect of the apparatus was compared[129] with data obtained with capillary tubes[56,60] using spores of *B. stearothermophilus* TH24 suspended in milk and water. In the steriliser, holding time distributions were determined by injecting a salt solution into the product flow and measuring concentrations at the end of the holding tube by an electrode. This method could not be used for milk because of its naturally high conductivity. It was found that the bacteriologically effective mean holding time was different from the mean holding time. At 145 °C the mean holding time was 3·26 secs while the bacteriologically effective mean holding time was 2·31 secs. This was essentially because those spores which travelled fastest through the steriliser had least exposure to the heat treatment and therefore a greater proportion of survivors. The bacterio-logically effective mean holding time was reduced as the temperature increased because of the decrease in resistance of the spores.

Thermal death of *B. stearothermophilus* TH24 spores followed Arrhenius kinetics over the temperature range 135–150 °C but when compared with the results from capillary tubes the UHT plant showed a greater change of rate of thermal death time with temperature than expected, the lines intersecting within a few degrees of 145 °C. The difference was less pronounced when water was used as the suspending medium. There appeared to be no explanation for the difference in results from the two methods of heating.

Using a similar laboratory scale steriliser, Neaves and Jarvis[90,91] studied the death rate of *Cl. botulinum* at high temperatures. In contrast to the results of *B. stearothermophilus* just described they found that the *D* values for *Cl. botulinum* were lower in the steriliser than in capillary tubes. The difference decreased with increase in temperature.

Steam into Product Methods
Dual purpose UHT steriliser. Burton and Perkin[130] described an experimental UHT steriliser capable of operating either as a plate type indirect heating system or as a direct steam into milk heating plant in which the two types of process could be compared. The heating profiles of the two systems differed in that the indirect system produced an exponential rate of heating whereas the steam injection method produced virtually instantaneous heating.

Milk inoculated with large numbers of spores of either *B. subtilis* 786 or *B. stearothermophilus* TH24 was heated using both methods in the UHT steriliser.[131] The temperature ranges employed were 135–142 °C for the indirect method and 132–147 °C for the direct method with mean holding times of 3·6 and 3·0 secs respectively. More reproducible results were obtained using *B. stearothermophilus* which showed that the direct heating system needed to be 3–4 °C higher to achieve the same sporicidal effect. The inhibitory action of UHT milk on survivor counts appeared to increase with increasing temperature and gave an apparent increase in kill of approximately 4 log reductions. The results indicated a Q_{10} of 29 for *B. stearothermophilus* in milk which was higher than the value of 11·5 obtained earlier using capillary tubes.[86]

Burton[132] investigated this difference in Q_{10} values and found that extrapolation of the capillary tube data[86] led to an overestimate of the sterilising effect of the direct heating UHT process. By calculation, the value for Q_{10} to give an accurate process estimation was 5·5. Using this value it was calculated that the thermal kill of the direct heating system at a temperature 3·5 °C higher would be equivalent to the indirect heating system. This had already been found experimentally.[131] It was concluded that the inhibitory effect of UHT milk was being carried over to the plating media leading to apparently high Q_{10} values being observed. This is obviously very important when calculating process requirements for milk in UHT systems if for some reason the inhibitory effect could not be relied upon.

Later experiments were reported[129] in which the dual purpose UHT steriliser in direct heating mode was compared with the laboratory scale product into steam UHT plant of Perkin,[128] using spores of *B. stearothermophilus*. Thermal death data from both sterilisers were very similar over the temperature range 135–150 °C. The bacteriologically effective mean holding time of the dual purpose steriliser decreased from around 3 secs at 130 °C to just under 2 secs at 150 °C.

Large scale direct steam injection system. Edwards *et al.*[133,134] used a large scale direct steam injection system operating between 112·8 °C and 135 °C to study the death rate of *B. subtilis* in skim milk. Holding times were measured using a salt solution and an automatic timer.

Significantly different results were obtained using two different recovery media, one containing calcium chloride and sodium dipicolinate and the other unsupplemented. The supplemented medium gave higher D values but lower z values than the unsupplemented medium. The thermal death time curves from both media were concave upwards with z values at the higher temperatures being twice as large (13·3 °C and 18·3 °C) as those at lower temperatures (6·7 °C and 8·9 °C).

It was also noted that precise temperature control was critical since small changes in temperature resulted in large changes in the number of survivors.

Busta[66] extended the work of Edwards *et al.*[133,134] to study the thermal death of spores of PA 3679 and *B. stearothermophilus* in the temperature ranges 115·6–132·2 °C and 126·7–137·8 °C respectively. Heat activation of *B. stearothermophilus* spores as well as inactivation was observed but no heat activation response was shown with PA 3679.

The D value for PA 3679 was less than 10 secs at 121·1 °C which is rather low for this organism but the z value was 17·8–19·4 °C which is high. The z values for *B. stearothermophilus* were 6·7 °C and 8·9 °C corresponding to different sporulation media. There was no evidence of curvature of the thermal death time curves for either organism in which respect they differed from the earlier work[133] on *B. subtilis*. It is possible that had a larger temperature range been covered, curvature may have been observed by using a thermal death time rather than an Arrhenius plot.

Alfa-Laval Vacu-Therm instant steriliser (VTIS). Lindgren and Swartling[135] heated spores of *B. subtilis* and *B. stearothermophilus* in milk in the VTIS plant operating at 130–140 °C by direct steam injection with a holding time of 3–4 secs. This was followed by partial evaporative cooling.

The sterilising effect[111] for *B. subtilis* varied from 6 at 120 °C to over 9 at 130 °C while for the more resistant *B. stearothermophilus* the sterilising effect at 120 °C was 4 and at 135 °C was over 7·4. The inhibitory effect of milk on survivor levels was not taken into account. In comparison, Franklin *et al.*[85,86] using the APV plate heater found that a temperature of 142 °C was necessary to achieve a sterilising effect of 7 for *B. stearothermophilus*, while a sterilising effect of 10 at 135 °C for *B. subtilis* was reported. Galesloot.[111] using the Stork tubular heater showed that 135 °C produced 6 log reductions of spores of *B. subtilis*. Different UHT milk sterilisers therefore produce comparable reductions in spore counts of both mesophilic and thermophilic sporeformers.[2]

Particle Heating Methods
It has been shown earlier in this chapter that process times for particles are considerably longer than those for liquids processed at the same temperature. It is therefore important to ensure that the process given is adequate to sterilise the particles to avoid spoilage of the product. In liquid/particle products, processes should ideally be determined for each system since the rates of heat transfer between the two will almost certainly vary from product to product.

Jacobs *et al.*[136] determined the death rate of *B. subtilis* spores located inside rolls of paper in glass tubes at temperatures between 104·4 °C and 132·2 °C. As expected, longer process times were required to achieve sterility in the tubes containing the inoculated rolls of paper than in the control tubes containing simply spores in broth.

The heat insulating effect of the paper was quantified by computing a protection ratio obtained by dividing the equivalent heating time (F_T) at temperature $T(F_T)$ required to sterilise the suspending medium by the equivalent processing time (θ_T) at temperature $T(\theta_T)$ at the coldest spot inside the paper roll when F_T had been achieved. In practice, the inclusion of the paper roll in the tube resulted in the process at 132·2 °C being increased 855 fold to effect sterilisation.

Perspex Beads
In this method reported by Hunter,[137] spores of *Bacillus anthracis* were embedded in 0·125 inch diameter particles of polymethylmethacrylate. The spores were mixed with liquid methacrylate, Tween 80 and powdered polymethylmethacrylate. The mixture was then polymerised in silicone rubber moulds at 49 °C for 6 h. After heat treatment in a direct steam injection system capable of operating between 104·4 °C and 154·4 °C, the

spores were recovered by dissolving the particles in acetone, filtering, incubating the filter on heart infusion agar and counting colonies.

Calibration curves were obtained at 90, 105 and 120 °C by heating spores with polymer in 0·6 mm bore capillary tubes in an oil bath. Death rate constants were calculated assuming both first order and Arrhenius kinetics. The results, when recalculated, suggest a z value of 61 °C for the system with D values at the three temperatures of 3·7, 2·1 and 1·2 mins. The z value is very high even for dry heat resistance where z values are generally of the order 14–27 °C.[41]

This method therefore appears to have limited application, since spore resistance characteristics in the 'perspex' are not at all comparable with spore resistance in food particles where much lower z values would be expected. The use of acetone in the recovery procedure may also be harmful to heat damaged spores.

Alginate Beads

Dallyn et al.[138] reported a method similar in principle to that of Hunter[137] but instead of 'perspex', the spores were immobilised in alginate gel, which provides moist heating conditions similar to food particles.

The immobilisation technique consisted of mixing spores with a solution of sodium alginate and dropping this into a solution of 2 % (w/v) calcium chloride using a hypodermic syringe. The drops solidified on contact with the calcium chloride as a result of the conversion of the soluble sodium alginate into insoluble calcium alginate. After curing from 18–48 h in the calcium chloride solution the beads were ready for use or alternatively could be stored at 4 °C for up to 12 months without deleterious effects on spore resistance or stability of the beads. Bead size was determined by the size of the syringe needle and the alginate strength. After processing the spores were released from the beads using 1 % potassium or sodium citrate.

It was found that viable spores could outgrow from the beads when they were placed in nutrient substances without prior dissolution of the alginate which would also allow estimation of process survival using a most probable number technique. Differences in resistance were also observed when the beads were heated in different food substrates.

Over the temperature range 120–145 °C, z values were determined[48] for spores of B. stearothermophilus suspended in buffer and in alginate beads. A linear relationship was observed between log D and temperature with a z value of 8·8 °C although the resistance of the spores in the alginate was lower than in the buffer solution. Above 145 °C the results were less consistent where D values were between 0·6 and 1 sec.

Alginate beads containing spores of *B. stearothermophilus* were passed successfully in a starch solution through a scraped surface UHT plant operating at 138, 140 and 142°C with two flow rates, 315/h and 450/h.[48,101] The mean residence time of 7·6 secs was maintained by adjusting the holding tube length. At 142°C with a flow rate of 315/h the sterilising effect was between 4 and 6 while at 138°C with a flow rate of 450/h less than 50% of the beads received a sterilising effect greater than 3.

Reconstituted Food Particles
The alginate bead technique[138] was modified by the authors[117,139,140] to incorporate spores into large particles of reformed food product. This was achieved by mixing together a purée of the product, sodium alginate, calcium sulphate, sodium citrate and spore suspension. The mixture was then allowed to set in moulds before use or storage for a short time at 4°C.

The D and z values of the spores of *B. stearothermophilus* and *Cl. sporogenes* (PA 3679) in the mixture were determined using the capillary tube technique. Meanwhile, the particles were heated in an experimental direct steam heating apparatus[101] in which the rate of heating at the centre of a particle could be monitored using a thin wire thermocouple. Spores were recovered from the particles by roughly chopping the particles and shaking in 5% sodium citrate solution. The recovery media were supplemented by the addition of 500 ppm Ca^{2+} to counteract the effect of sodium citrate carried over.

The experimental F_s value (F value integrated over the whole particle) was derived from the equation given by Stumbo:[15]

$$F_s = D_{121·1} (\log a - \log b)$$

where a = initial spore concentration, b = final spore concentration and $D_{121·1} = D$ value at 121·1°C. This value was then compared with the calculated F_s value from the following equation adapted from Stumbo:[15]

$$F_s = F_c + D_{121·1} \log \left(\frac{D_{121·1} + 11·7317(F_\lambda - F_c)}{D_{121·1}} \right)$$

where $F_c = F$ value at centre of particle, $F_\lambda = F$ value where the lag factor j = half the j value at the centre and 11·7317 is a constant. This equation assumes a linear z value, which if valid, should lead to similar values for F_s by each method. Figure 7 shows the results obtained for *B. stearothermophilus* spores heated in reformed meat cubes (1·8 cm) over the temperature range 125–150°C.[141] Using this method it should be possible to monitor

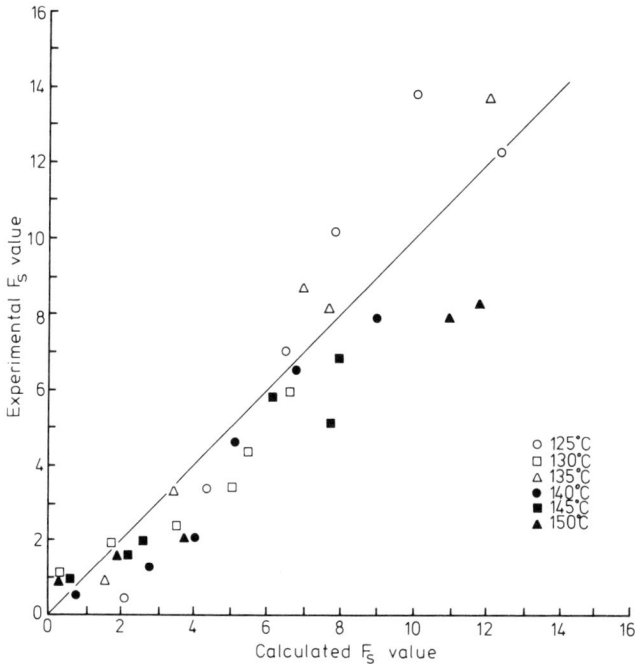

FIG. 7. Comparison of experimental and calculated F_s values for *Bacillus stearothermophilus* spores heated in 1·8 cm meat/alginate cubes over the temperature range 125–150 °C.[141]

spore destruction in particles of food of any shape or size in UHT particle sterilisation systems.

ACKNOWLEDGEMENT

The authors wish to thank the Director of the Campden Food Preservation Research Association for permission to publish this article.

REFERENCES

1. BALL, C. O., and OLSON, F. C. W., *Sterilization in Food Technology*, 1957, McGraw-Hill Book Co. Inc., New York.
2. BURTON, H., *J. Soc. Dairy Technol.*, 1965, **18**, 58.
3. SACHAROW, S., *Fd Trade Rev.*, 1973, **43**, 19.

4. BALL, C. O., *Fd Research*, 1938, **3**, 13.
5. STUMBO, C. R., *Proc. 2nd Int. Congr. Fd Sci. and Technol.*, Warszawa, Poland, 1966, 171.
6. ANON, *Packaging Rev.*, 1972, **92**, 37.
7. ANON, *Fd Engng*, 1953, **25**, 102.
8. HAVIGHORST, C. R., *Fd Engng*, 1953, **25**, 76.
9. HERSOM, A., *Fd Manufacture*, **44**, 28.
10. LAWLER, F. K., *Fd Engng*, 1967, **39**, 58.
11. BALL, C. O., JOFFE, F. M., STIER, E. F., and HAYAKAWA, K., *ASHRAE J.*, 1963, **5**, 93.
12. ANON, *Fd Process. Industry*, 1970, **39**, 37.
13. HORNER, S. A., WALLEN, S. E., and CAPORASO, F., *J. Fd Prot.*, 1980, **43**, 54.
14. MEHTA, R. S., *J. Fd Prot.*, 1980, **43**, 212.
15. STUMBO, C. R., *Thermobacteriology in Food Processing*, 1965, Academic Press, New York and London.
16. SHORE, D. T., *Ultra-High-Temperature Processing of Dairy Products*, 1970, Society of Dairy Technology, London.
17. NEWMAN, M. E., and STEELE, D. A., *Technical Memorandum No. 210*, 1978, Campden Food Preservation Research Association, Chipping Campden, England.
18. CHARLETTE, S. M., *Fd Trade Rev.*, 1971, **41**, 25.
19. BALL, R., *J. Soc. Dairy Technol.*, 1977, **30**, 143.
20. SHORE, T., *Proc. Biochem.*, 1966, **1**, 103.
21. ANON, *Fd Engng*, 1953, **25**, 74.
22. NORTHCUTT, R. T. JR., and GEMMILL, A. V., *Fd Engng*, 1957, **29**, 66.
23. PERKIN, A. G., and BURTON, H., *J. Soc. Dairy Technol.*, 1970, **23**, 147.
24. DEINDOERFER, F. H., *Appl. Microbiol.*, 1957, **5**, 221.
25. DEINDOERFER, F. H., and HUMPHREY, A. E., *Appl. Microbiol.*, 1959, **7**, 264.
26. ANON., *Fd Processing*, 1963, **24**, 81.
27. BURTON, H., *J. Dairy Res.*, 1958, **25**, 75.
28. BURTON, H., *J. Dairy Res.*, 1958, **25**, 324.
29. BATESON, R. N., *Chem. Engng Progr. Symp. Ser. No. 108*, 1971, **67**, 44.
30. RAO, M. A., and LONCIN, M., *Lebensm.-Wiss. u. Technol.*, 1974, **7**, 5.
31. RAO, M. A., and LONCIN, M., *Lebensm.-Wiss. u. Technol.*, 1974, **7**, 14.
32. ROIG, S. M., VITALI, A. A., RODRIGUEZ, E. O., and RAO, M. A., *Lebensm.-Wiss. u. Technol.*, 1976, **9**, 255.
33. KAUFMAN, U. F., PUTNAM, G. W., and IJICHI, K., *J. Milk and Fd Technol.*, 1968, **31**, 310.
34. CHARM, S. E., *Fd Technol.*, 1966, **20**, 665.
35. SPECK, M. L., and BUSTA, F. F., *J. Dairy Sci.*, 1968, **51**, 1146.
36. BURTON, H., *Ultra-High-Temperature Processing of Dairy Products*, 1970, Society of Dairy Technology, London.
37. FRANKLIN, J. G., WILLIAMS, D. J., and CLEGG, L. F. L., *J. Appl. Bact.*, 1956, **19**, 46.
38. ASHTON, T. R., *Ultra-High-Temperature Processing of Dairy Products*, 1970, Society of Dairy Technology, London.
39. LUCK, H., MOSTERT, J. F., and HUSMANN, R. A., *S. Afr. J. Dairy Technol.*, 1978, **10**, 83.

40. PFLUG, I. J., and SCHMIDT, C. F., *Disinfection, Sterilization and Preservation*, eds. Lawrence, C. A. and Block, S. S., 1968, Henry Kimpton, London.
41. RUSSELL, A. D., *Inhibition and Destruction of the Microbial Cell*, ed. Hugo, W. B., 1971, Academic Press, London.
42. ROBERTS, T. A., and HITCHINS, A. D., *The Bacterial Spore*, eds. Gould, G. W. and Hurst, A., 1969, Academic Press, London.
43. MURRELL, W. G., *Flame Sterilization. Specialist Courses in the Food Industry No. 2.* 1972, AIFST–CSIRO, 12.
44. WILLIAMS, C. C., MERRILL, C. M., and CAMERON, E. J., *Fd Research*, 1937, **2**, 369.
45. PFLUG, I. J., and SMITH, G. M., *Spore Research 1976* vol. II, eds. Barker, A. N., Wolf, J., Ellar, D. J., Dring, G. J., and Gould, G. W. 1977, Academic Press, London.
46. SHULL, J. J., and ERNST, R. R., *Appl. Microbiol.*, 1962, **10**, 452.
47. LEE, Y. H., Growth, sporulation and spore properties of *Bacillus stearothermophilus* produced in chemically defined media. Ph.D. Thesis, 1976, University of Aston, Birmingham, England.
48. BEAN, P. G., DALLYN, H., and RANJITH, H. M. P., *Food Microbiology and Technology*, eds. Jarvis, B., Christian, J. H. B., and Michener, H. D., 1979, Medicina Viva Servizio Congressi, Srl, Parma, Italy.
49. MOATS, W. A., DABBAH, R., and EDWARDS, V. M., *J. Fd Sci.*, 1971, **36**, 523.
50. MOATS, W. A., *J. Bact.*, 1971, **105**, 165.
51. HUMPHREY, A. E., and NICKERSON, J. T. R., *Appl. Microbiol.*, 1961, **9**, 282.
52. CHARM, S. E., *Fd Technol.*, 1958, **12**, 4.
53. AIBA, S., and TODA, K., *Proc. Biochem.*, 1967, **2**, 35.
54. RUSSELL, A. D., *Mfg Chemist and Aerosol News*, 1965, **36**, 38.
55. SHULL, J. J., CARGO, G. T., and ERNST, R. R., *Appl. Microbiol.*, 1963, **11**, 485.
56. PERKIN, A. G., BURTON, H., UNDERWOOD, H., and DAVIES, F. L., *J. Fd Technol.*, 1977, **12**, 131.
57. GILLESPY, T. G., *A. Rep. Fruit Veg. Presn Res. Stn*, 1948, Chipping Campden, Glos., England.
58. ESTY, J. R., and MEYER, K F, *J. Infect. Dis.*, 1922, **31**, 650.
59. WANG, D.I-C., SCHARER, J., and HUMPHREY, A. E., *Appl. Microbiol.*, 1964, **12**, 451.
60. DAVIES, F. L., UNDERWOOD, H. M., PERKIN, A. G., and BURTON, H., *J. Fd Technol.*, 1977, **12**, 115.
61. NAVANI, S. K., SCHOLEFIELD, J., and KIBBY, M. R., *J. Appl. Bact.*, 1970, **33**, 609.
62. ESSELEN, W. B., and PFLUG, I. J., *Fd Technol.*, 1956, **10**, 557.
63. LICCIARDELLO, J. J., and NICKERSON, J. T. R., *Appl. Microbiol.*, 1963, **11**, 476.
64. REICHART, O., *Acta Alimentaria*, 1979, **8**, 131.
65. COWELL, N. D., *J. Fd Technol.*, 1968, **3**, 303.
66. BUSTA, F. F., *Appl. Microbiol.*, 1967, **15**, 640.
67. COOK, A. M., and GILBERT, R. J., *J. Fd Technol.*, 1968, **3**, 295.
68. AMAHA, M. and ORDAL, Z. J., *J. Bact.*, 1957, **74**, 569.
69. SUGIYAMA, H., *J. Bact.*, 1951, **62**, 81.
70. FIELDS, M. L., and FINLEY, N., Research Bull. 807, 1962, Univ. Missouri College of Agric., USA.

71. COOK, A. M., and GILBERT, R. J., *J. Pharm. Pharmacol.*, 1968, **20**, 626.
72. CERNY, G., *Z. Lebensm. Unters. Forsch.*, 1980, **170**, 180.
73. BROWN, K. L., and THORPE, R. H., Technical Memorandum No. 185, 1978, Campden Food Preservation Research Association, Chipping Campden, Glos., England.
74. BROWN, K. L., and THORPE, R. H., Technical Memorandum No. 204, 1978, Campden Food Preservation Research Association, Chipping Campden, Glos., England.
75. AMAHA, M., and SAKAGUCHI, K. I., *J. Bact.*, 1954, **68**, 338.
76. COOK, A. M., and GILBERT, R. J., *J. Fd Technol.*, 1968, **3**, 285.
77. FRANK, H. A., and CAMPBELL, L., *Appl. Microbiol.*, 1955, **3**, 300.
78. OLSEN, A. M., and SCOTT, W. J., *Aust. J. Sci. Res.*, 1950, **B3**, 219.
79. PRENTICE, G. A., and CLEGG, L. F. L., *J. Appl. Bact.*, 1974, **37**, 501.
80. OQUENDO, R., VALDIVIESO, L., STAHL, R., and LONCIN, M., *Lebensm.-Wiss. u. Technol.*, 1975, **8**, 181.
81. BIGELOW, W. D., and ESTY, J. R., *J. Infect. Dis.*, 1920, **27**, 602.
82. ESTY, J. R., and WILLIAMS, C. C., *J. Infect. Dis.*, 1924, **34**, 516.
83. WILDER, C. J., and NORDAN, H. C., *Fd Research*, 1957, **22**, 462.
84. FRANKLIN, J. G., WILLIAMS, D. J., and CLEGG, L. F. L., *J. Appl. Bact.*, 1958, **21**, 51.
85. FRANKLIN, J. G., WILLIAMS, D. J., CHAPMAN, H. R., and CLEGG, L. F. L., *J. Appl. Bact.*, 1958, **21**, 47.
86. FRANKLIN, J. G., WILLIAMS, D. J., BURTON, H., CHAPMAN, H. R., and CLEGG, L. F. L., *Int. Dairy Congress 15*, vol. 1, 1959, London.
87. RESENDE, R., Regeneration of thermally inactivated enzymes in pH adjusted chlorophyll-containing vegetables processed by HTST methods, Ph.D. Thesis, 1966, University of Massachusetts, Amherst, USA.
88. RESENDE, R., STUMBO, C. R., and FRANCIS, F. J., *Fd Technol.*, 1969, **23**, 325.
89. OLSON, F. C. W., and SCHULTZ, O. T., *Ind. Engng Chem.*, 1942, **34**, 874.
90. NEAVES, P. and JARVIS, B., Research Report No. 280, 1978, BFMIRA, Leatherhead, Surrey, England.
91. NEAVES, P., and JARVIS, B., Research Report No. 286, 1978, BFMIRA, Leatherhead, Surrey, England.
92. STERN, J. A., and PROCTOR, B. E., *Fd Technol.*, 1954, **8**, 139.
93. STERN, J. A., HERLIN, M. A., and PROCTOR, B. E., *Fd Research*, 1952, **17**, 460.
94. OLSON, F. C. W., and JACKSON, J. M., *Ind. Engng Chem.*, 1942, **34**, 337.
95. CERF, O., GROSCLAUDE, G., and VERMIERE, D., *Appl. Microbiol.*, 1970, **19**, 696.
96. BROWN, K. L., *Aseptic Packaging of Vegetable Products*, 1974, Campden Food Preservation Research Association, Chipping Campden, Glos., England.
97. BROWN, K. L., *Aseptic Packaging of Vegetable Products*, 1975, Campden Food Preservation Research Association, Chipping Campden, Glos., England.
98. REICHART, O., *Acta Alimentaria*, 1978, **7**, 396.
99. MICHIELS, L., *Industr. Alim. Agr.*, 1972, **89**, 1349.
100. THORPE, R. H., and GREY, K. A., Technical Note 138, 1971, Campden Food Preservation Research Association, Chipping Campden, Glos., England.

101. PERKIN, A. G., DAVIES, F. L., NEAVES, P., JARVIS, B., AYRES, C. A., BROWN, K. L., FALLOON, W. C., DALLYN, H., and BEAN, P., *Microbial Growth and Survival in Extremes of Environment*, eds. Gould, G. W. and Corry, J. E. L. 1980, Academic Press, London.
102. PFLUG, I. J., SMITH, G., HOLCOMB, R., and BLANCHETT, R., *J. Fd Prot.*, 1980, **43**, 119.
103. TOWNSEND, C. T., ESTY, J. R., and BASELT, F. C., *Fd Research*, 1938, **3**, 323.
104. SOGNEFEST, P., HAYS, G. L., WHEATON, E., and BENJAMIN, H. A., *Fd Research*, 1948, **13**, 400.
105. REYNOLDS, H., KAPLAN, A. M., SPENCER, F. B., and LICHTENSTEIN, H., *Fd Research*, 1952, **17**, 153.
106. SCHMIDT, C. F., *J. Bact.*, 1950, **59**, 433.
107. RICHARDSON, A. C., Unpublished data, 1926.
108. BALL, C. O., MERRILL, C. M., WILLIAMS, C. C., and WESSEL, D. J., Unpublished data, 1937.
109. SOGNEFEST, P., and BENJAMIN, H. A., *Fd Research*, 1944, **9**, 234.
110. CUNCLIFFE, H. R., BLACKWELL, J. H., DORS, R., and WALKER, J. S., *J. Fd Prot.*, 1979, **42**, 135.
111. GALESLOOT, TH. E., *Neth. Milk Dairy J.*, 1956, **10**, 79.
112. MARTIN, W. McK., *Canner*, 1950, **111**, 12.
113. MARTIN, W. McK., *Martin Aseptic Canning System*, 1950, Decennial IFT Conference, Chicago.
114. WILLIAMS, D. J., FRANKLIN, J. G., CHAPMAN, H. R., and CLEGG, L. F. L., *J. Appl. Bact.*, 1957, **20**, 43.
115. STROUP, W. H., DICKERSON, R. W., JR., and READ, R. B. JR., *Appl. Microbiol.*, 1969, **18**, 889.
116. KOOIMAN, W. J., *Spore Research 1973* eds. Barker, A. N., Gould, G. W., and Wolf, J., 1974, Academic Press, London.
117. BROWN, K. L., and AYRES, C. A., Technical Memorandum No. 186, 1978, Campden Food Preservation Research Association, Chipping Campden, Glos., England.
118. STUMBO, C. R., *Fd Technol.*, 1948, **2**, 228.
119. PFLUG, I. J., and ESSELEN, W. B., *Fd Technol.*, 1953, **7**, 237.
120. PFLUG, I. J., and ESSELEN, W. B., *Agric. Engng*, 1954, **35**, 245.
121. PFLUG, I. J., *Fd Technol.*, 1960, **14**, 483.
122. STUMBO, C. R., MURPHY, J. R., and COCHRAN, J., *Fd Technol.*, 1950, **4**, 321.
123. PFLUG, I. J., and ESSELEN, W. B., *Fd Research*, 1954, **19**, 92.
124. SECRIST, J. L., and STUMBO, C. R., *Fd Research*, 1958, **23**, 51.
125. SECRIST, J. L., and STUMBO, C. R., *Fd Technol.*, 1956, **10**, 543.
126. FRANK, H. A., and CAMPBELL, L. L. JR., *Appl. Microbiol.*, 1957, **5**, 243.
127. YOULAND, G. C., and STUMBO, C. R., *Fd Technol.*, 1953, **7**, 286.
128. PERKIN, A. G., *J. Dairy Res.*, 1974, **41**, 55.
129. BURTON, H., PERKIN, A. G., DAVIES, F. L., and UNDERWOOD, H. M., *J. Fd Technol.*, 1977, **12**, 149.
130. BURTON, H., and PERKIN, A. G., *J. Dairy Res.*, 1970, **37**, 209.
131. FRANKLIN, J. G., UNDERWOOD, H. M., PERKIN, A. G., and BURTON, H., *J. Dairy Res.*, 1970, **37**, 219.
132. BURTON, H., *J. Dairy Res.*, 1970, **37**, 227.

133. EDWARDS, J. L. JR., BUSTA, F. F., and SPECK, M. L., *Appl. Microbiol.*, 1965, **13**, 851.
134. EDWARDS, J. L. JR., BUSTA, F. F., and SPECK, M. L., *Appl. Microbiol.*, 1965, **13**, 858.
135. LINDGREN, B., and SWARTLING, P., *Milk and Dairy Res.*, Rep. No. 69, 1963, Alnarp.
136. JACOBS, R. A., KEMPE, L. L., and MILONE, N. A., *J. Fd Sci.*, 1973, **38**, 168.
137. HUNTER, G. M., *Fd Technol. in Australia*, 1972, **24**, 158.
138. DALLYN, H., FALLOON, W. C., and BEAN, P. G., *Lab. Practice*, 1977, **26**, 773.
139. BROWN, K. L., and AYRES, C. A., Technical Memorandum No. 203, 1978, Campden Food Preservation Research Association, Chipping Campden, Glos., England.
140. AYRES, C. A., and BROWN, K. L., Technical Memorandum No. 224, 1979, Campden Food Preservation Research Association, Chipping Campden, Glos., England.
141. AYRES, C. A., and BROWN, K. L., Technical Memorandum No. 250, 1980, Campden Food Preservation Research Association, Chipping Campden, Glos., England.

Chapter 5

REGULATION OF LACTOSE METABOLISM IN DAIRY STREPTOCOCCI

L. L. McKay

Department of Food Science and Nutrition,
University of Minnesota, USA

SUMMARY

The economic value of dairy streptococci depends in part on their efficient fermentation of lactose to lactic acid. The genetic and metabolic regulation of this process has only recently received intense research investigation. The initial steps of lactose utilisation may involve galactoside permease and β-galactosidase as in Streptococcus thermophilus *or a phosphoenolpyruvate (PEP)-dependent phosphotransferase system (PTS) coupled with phospho-β-galactosidase as exists in group N streptococci. Both systems may occur simultaneously in some strains. The instability of lactose-fermenting ability observed in many group N streptococci is linked to plasmid DNA carrying the* lac *genes. The acquisition of this plasmid may have allowed these organisms to adapt to milk as a natural ecological niche. Stabilisation of lactose metabolism can be accomplished by inserting the Lac plasmid, or a portion of it, into the host chromosome. Models have been proposed to account for the appearance of partial lactose-fermenting revertants from lactose-negative clones of lactic streptococci. Our knowledge concerning the regulation of lactose metabolism in dairy streptococci is by no means complete. Further research in this area should eventually lead to the development of improved lactose-fermenting strains.*

INTRODUCTION

Dairy streptococci, defined herein as strains of *Streptococcus lactis*, *S. cremoris*, *S. lactis* subsp. *diacetylactis*, and *S. thermophilus*, have been utilised for centuries in the production of fermented dairy products. The manufacture of these products requires that the strains, with the possible exception of *S. diacetylactis*, convert the lactose in milk to lactic acid.

When the streptococci are added to a medium such as milk, containing lactose as the primary carbohydrate source, they must hydrolyse the sugar to monosaccharides which then enter the glycolytic pathway. Through this scheme the organism acquires the necessary energy for growth and produces essential metabolites required for cell synthesis, along with copious amounts of lactic acid. Lactic acid contributes to the product's flavour and texture and aids in preventing the growth of spoilage and pathogenic microorganisms. Control of this process has only recently received extensive research investigation, especially as related to the enzymatic and genetic regulation of lactose metabolism. This is somewhat surprising when one considers the economic value of dairy products. For example, in 1976 fermented dairy products accounted for about 21·6% of the total value of fermented foods produced throughout the world.[76]

It is the intent of this chapter to describe recent developments in the genetic and enzymatic regulation of lactose metabolism in dairy streptococci. Pertinent to this discussion are the mechanisms used by dairy streptococci to hydrolyse lactose and the instability of lactose fermentation in the group N streptococci (*S. lactis*, *S. cremoris*, and *S. diacetylactis*).

BIOCHEMICAL MECHANISMS OF LACTOSE UTILISATION

Microorganisms utilise lactose by two principal methods. It is either hydrolysed by β-galactosidase (β-Gal) to glucose and galactose or lactose phosphate is formed and subsequently hydrolysed by phospho-β-galactoside galactohydrolase (P-β-Gal) to glucose and galactose-6-phosphate. Knowledge regarding the metabolism of lactose by both systems stems from extensive investigations conducted in *Escherichia coli* and *Staphylococcus aureus*. These systems are briefly summarised, since streptococci used in dairy fermentations utilise lactose by both mechanisms.

Lactose Utilisation in Escherichia coli
In *E. coli* a cluster of three genes constitutes the lactose operon. These

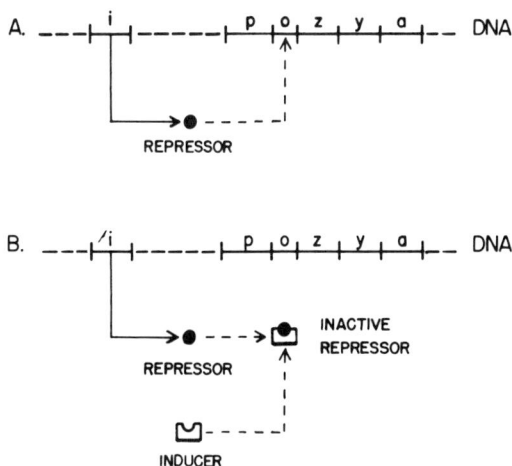

FIG. 1.   The lactose operon of *E. coli*. The inactive form of the operon (A) results because the regulator gene '*i*' synthesises a repressor protein which binds to the operator region and prevents synthesis of mRNA for β-Gal and galactoside permease. The active form of the operon (B) results when an inducer is present which binds to the repressor converting it to an inactive form. This allows synthesis of the lac-specific proteins.

include the '*z*' gene which codes for the structure of β-Gal, the '*y*' gene which directs the synthesis of a permease required for an energy-dependent active transport of lactose into the cell,[36] and the '*a*' gene which is not essential for growth or lactose metabolism. These three genes are transcribed into a single polycistronic messenger RNA (mRNA) molecule, and are thus said to constitute an operon (Fig. 1).

The synthesis of lac mRNA is controlled by a protein molecule termed the repressor and is coded for by the '*i*' gene. Genes of this type are also termed regulatory genes. The repressor binds to the operator '*o*' region of the lac operon and blocks the transcription of lac mRNA. The synthesis of the lac-specific enzymes is thus said to be repressed. When the repressor is inactivated, either by mutation or binding reversibly to small molecules (inducers), it falls off the '*o*' locus allowing synthesis of the lac enzymes. The operon is now said to be induced. The repressor has two active sites, one which interacts with the operator and the other which interacts with the inducer. Mutations in the former site, which prevent binding of the repressor to the operator ($i^+ \rightarrow i^-$) result in maximal synthesis of the lac proteins even in the absence of inducer (constitutive mutant) whereas mutations in the latter site would exist in a continuously repressed state

(Lac⁻) regardless of the presence of inducer. These mutants, termed i^s, are thought to produce a repressor which has no affinity for inducer; subsequently, the operon is permanently shut off.[4] Mutations in the 'o' gene, in which the operator region is altered such that the repressor can no longer bind with the operator, are termed o^c and cause constitutive expression of the *lac* genes linked to the operator.[4] Lac⁻ mutants could also occur as a result of mutations in z ($z^+ \rightarrow z^-$) or in y ($y^+ \rightarrow y^-$). Beckwith[4] introduced the term promoter (p) to indicate the site on the genome where synthesis of messenger RNA for the lac operon begins. This site sets the maximal potential for expression of the lac operon and promoter mutants from *E. coli* with an altered maximal level of the lac enzymes were isolated.

Lactose Utilisation in Staphylococcus aureus

In contrast to *E. coli* in which lactose is transported into the cell without modification, the transport of this disaccharide in *S. aureus* is achieved by group translocation, i.e. lactose is phosphorylated to lactose-6-phosphate during the transport process. The components responsible for this translocating process are arranged to catalyse a vectorial phosphorylation that transports and accumulates lactose into the cell interior where it is found as the phosphorylated derivative.[19]

Lactose transport and metabolism in *S. aureus* were reviewed by Hengstenberg *et al.*[22] The enzyme system performing the vectorial phosphorylation is the phosphoenolpyruvate (PEP)-dependent phosphotransferase system (PTS), which has been detected in many different microorganisms.[70] The lactose–PTS reaction scheme in *S. aureus* is described as follows:[70]

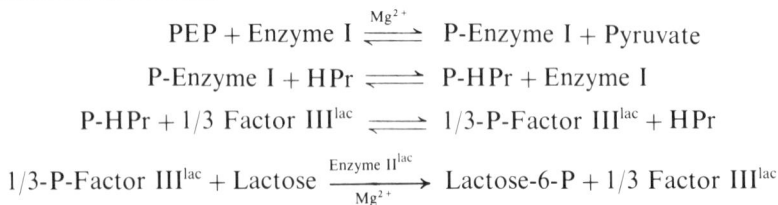

$$\text{PEP} + \text{Enzyme I} \underset{}{\overset{Mg^{2+}}{\rightleftharpoons}} \text{P-Enzyme I} + \text{Pyruvate}$$

$$\text{P-Enzyme I} + \text{HPr} \rightleftharpoons \text{P-HPr} + \text{Enzyme I}$$

$$\text{P-HPr} + 1/3 \text{ Factor III}^{lac} \rightleftharpoons 1/3\text{-P-Factor III}^{lac} + \text{HPr}$$

$$1/3\text{-P-Factor III}^{lac} + \text{Lactose} \xrightarrow[Mg^{2+}]{\text{Enzyme II}^{lac}} \text{Lactose-6-P} + 1/3 \text{ Factor III}^{lac}$$

Enzyme I and HPr are soluble cytoplasmic protein components that are synthesised constitutively and are necessary for the transport of many sugars. Factor III^{lac} is a soluble cytoplasmic galactoside-specific component and Enzyme II^{lac} is a membrane-bound component which is galactoside-specific.[22] The two lactose-specific proteins, Factor III^{lac} and Enzyme II^{lac}, are formed upon induction of the staphylococcal lac operon, as is P-β-Gal, the enzyme which cleaves lactose-6-P to galactose-6-P and glucose.[21,61]

The natural inducer for this lac operon is galactose-6-P.[22] Galactose only functions as an inducer in PTS wild type strains and lactose serves as the inducer only if the P-β-Gal and the PTS components are present. If the PTS were missing, the induction would only occur with galactose-6-P.

Since lactose is utilised via the PEP–PTS in *S. aureus*, several types of mutations can be described that affect its metabolism. Deficiency in Enzyme I or HPr results in the inability to utilise several carbohydrates, including lactose.[12,78] A defect in Enzyme II^{lac} gives rise to a Lac⁻ Gal⁻ phenotype as does a defect in Factor III^{lac}.[21,61,78] Strains unable to hydrolyse lactose-6-P are designated z^- and are missing P-β-Gal,[61] hence, they possess a Lac⁻ Gal⁺ phenotype. In addition, regulatory mutants (r^-) that are constitutive for the uptake and hydrolysis of lactose have been isolated. These different mutant types suggest that the lac operon of *S. aureus* could be represented as $r^+ z^+ II^+ III^+$. Transductional analysis has shown that z, II, and III are closely linked. It is not known whether a mutation in the r gene is comparable to i^- or o^c mutations in *E. coli*.[61]

Lactose Metabolism in Group N Streptococci
The lactic streptococci possess at least two mechanisms for the utilisation of lactose (Fig. 2). Citti *et al.*[5] showed that *S. lactis* 7962 possesses a typical β-Gal which suggests that lactose metabolism in this strain occurs as in *E. coli*. Kashkett and Wilson[31] confirmed this by noting that the non-metabolisable lactose analogue, thiomethylgalactoside (TMG), accumulated in 7962 in the unphosphorylated form. In addition, McKay *et al.*[57] examined the nature of the defective function in Lac⁻ mutants of 7962 by looking at β-Gal activity and [^{14}C] TMG uptake, which is an indicator of β-galactoside permease activity. These mutants were separated into several classes such as $z^+ y^-$, $z^- y^-$ (this mutant was also Gal⁻) and $z^- y^+$. Although lactose was utilised, the latter mutant could not hydrolyse o-nitrophenyl-β-D-galactopyranoside (ONPG, a chromogenic substrate for β-Gal), which suggested the β-Gal had lost affinity for ONPG but not for the natural substrate lactose. The regulatory control of lactose metabolism in *S. lactis* 7962 appeared to be similar but not identical to that in *E. coli*.

In 1965, Citi *et al.*[5] reported that induced whole cells of group N streptococci hydrolysed ONPG, yet of more than 40 strains of *S. lactis*, *S. cremoris*, and *S. diacetylactis* examined, only *S. lactis* 7962 possessed β-Gal activity in toluene-treated cells or cell-free extracts. Attempts to stabilise the enzyme in the other strains using conventional enzyme procedures were unsuccessful and it was concluded that these strains possessed a 'labile' β-Gal. The apparent instability of this enzyme was subsequently explained by

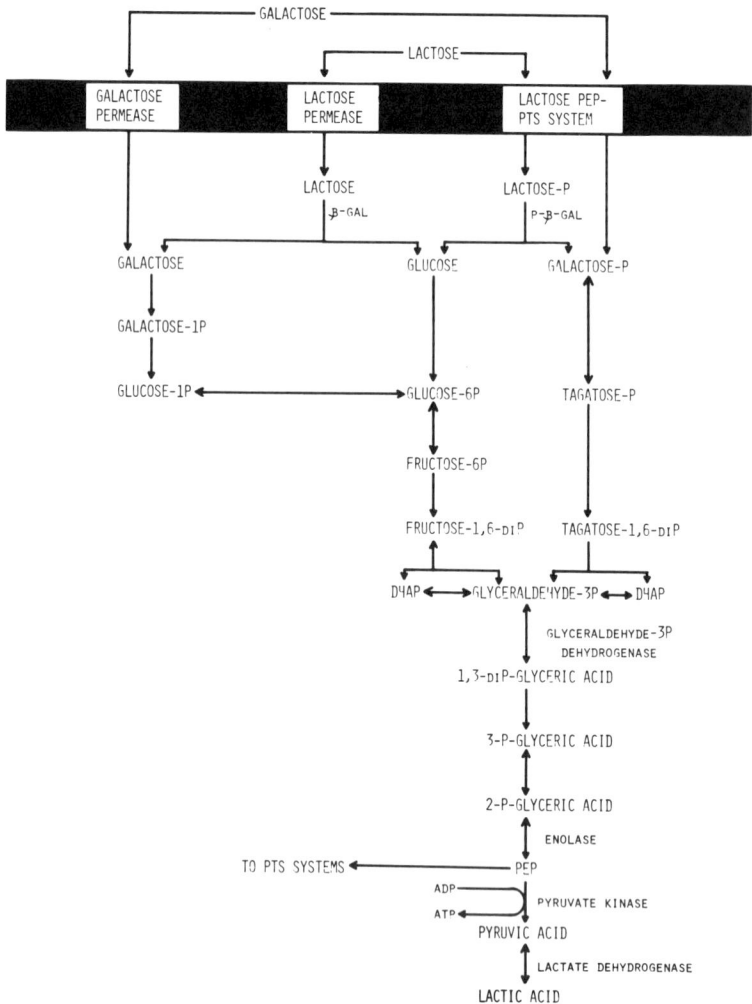

FIG. 2.   Pathways for lactose and galactose catabolism in dairy streptococci.
A strain may possess one or more of the indicated transport mechanisms.

the fact that these organisms did not possess β-Gal.[57,58] Instead, it was
shown that toluene-treated cells or cell-free extracts could not hydrolyse
lactose or ONPG but did possess an enzyme (P-β-Gal) which hydrolysed
the phosphorylated derivatives of these compounds as formed by a
PEP–PTS similar to that described for *S. aureus.*

The importance of PEP in lactose metabolism was first suggested by the observation that sodium fluoride (NaF) prevented lactose utilisation and ONPG hydrolysis in whole cells of *S. lactis* C2.[58] NaF prevented PEP generation during glycolysis by inhibiting enolase (Fig. 2). Furthermore, only the addition of PEP to toluene-treated cells of *S. lactis* C2 restored the ability to hydrolyse ONPG. The use of concentrated cell-free extracts from this organism revealed that ONPG hydrolysis was stimulated by the addition of PEP. NaF inhibited the hydrolysis of ONPG, yet the addition of PEP in the presence of NaF restored maximum activity. Addition of other high energy compounds such as acetyl phosphate, carbamyl phosphate, adenosine-5′-triphosphate, guanosine-5′-triphosphate, or uridine-5′-triphosphate did not replace the PEP requirements.

Experiments testing for the presence of phosphorylated lactose in *S. lactis* C2 using [^{14}C] lactose were unsuccessful due to rapid metabolism of the sugar and the absence of a mutant which would accumulate lactose or lactose-P without subsequent hydrolysis.[58] However, when TMG, the non-metabolisable analogue of lactose was used, the derivative TMG-P was shown to accumulate. This finding supported the hypothesis that lactose fermentation in C2 required substrate phosphorylation by a PEP-dependent system.

The importance of PEP in lactose metabolism was substantiated by a series of elegant experiments by Thompson and Thomas[85] and by Thomas.[83,84] Using starved cells of *S. lactis* ML$_3$, these workers were able to regulate, *in vivo*, the formation of PEP from glycolysis. The starved cells were devoid of glucose-6-phosphate, fructose-6-phosphate, fructose-1-6-diphosphate and triose phosphates but contained an intracellular pool of the three glycolytic intermediates, 3-phosphoglycerate (28·9 mM; 3-PG), 2-phosphoglycerate (5·3 mM; 2-PG) and PEP (11·3 mM) which were equivalent to a PEP-generating potential of approximately 45 mM.[83] This PEP-generating potential was subsequently shown to be the endogenous energy source for TMG accumulation by the starved cells, based on the following observations: (1) the intracellular concentration of PEP, 2-PG, and 3-PG decreased with concomitant uptake of TMG; (2) TMG accumulated as a phosphate derivative; (3) sodium fluoride inhibition of enolase prevented the conversion of 2-PG to PEP and uptake of TMG by the starved cells was reduced by 80%; and (4) the stoichiometric ratio, TMG accumulated/PEP consumed, was almost unity.[85]

Knowledge that starved cells of *S. lactis* ML$_3$ maintained an intracellular PEP potential and finding that the PTS was resistant to inhibition by iodoacetate (IAA), enabled Thompson[84] to isolate the first intracellular derivative of the PEP-dependent lactose transport system in *S. lactis* ML$_3$.

When accumulation of [¹⁴C]lactose by IAA-inhibited, starved cells was stopped within 1 s of commencement of transport, a phosphorylated disaccharide was identified. The compound was isolated and enzymatic analysis showed the derivative to be lactose-6-phosphate. Lee *et al.*[44] also provided evidence for the phosphorylation of the galactose moiety of lactose during its transport by means of the PTS with PEP as the phosphate donor. Thus, the importance of PEP in lactose metabolism by group N streptococci has been established.

As previously noted, ONPG hydrolysis could not be demonstrated in toluene-treated cells or cell-free extracts of *S. lactis* C2. However, such preparations did hydrolyse the substrate if PEP was added to the reaction mixture.[58] This result suggested that the proper substrate was ONPG-P and not ONPG, as phosphorylation of ONPG by the PEP-dependent system could occur in the presence of PEP. When various strains of *S. lactis*, *S. cremoris*, and *S. diacetylactis* were screened for their ability to hydrolyse ONPG-P or ONPG in toluene-treated cells, all but one were shown to hydrolyse ONPG-P and not ONPG.[57] The only exception was *S. lactis* 7962 which hydrolysed ONPG but not ONPG-P. Therefore, instead of having β-Gal, cells possessed P-β-Gal, which hydrolysed the phosphorylated derivatives of lactose or ONPG. This P-β-Gal has subsequently been purified and characterised[30,59] and appears to be analogous to the one found in *S. aureus*.[19,20]

Group N streptococcal utilisation of lactose via PTS components was verified by McKay *et al.*[57] who demonstrated that *S. lactis* C2 contained Enzyme I, HPr, Enzyme II^lac, and Factor III^lac. Complementation tests were performed in which cell extracts from *S. aureus* mutants missing each of these components were mixed with dilute cell-free extracts from *S. lactis* C2. No ONPG hydrolysis occurred in the test system unless the C2 extracts contained the component missing in the *S. aureus* mutant. When C2 contained the missing protein, the system was completed and ONPG hydrolysis restored. Not only did *S. lactis* C2 contain the PTS proteins, but they were interchangeable with the *S. aureus* components.[57] Thus, components of the PEP–PTS for lactose metabolism in group N streptococci appear identical to that reported for *S. aureus*.[22]

Lac⁻ mutants from *S. lactis* C2 were also isolated to further characterise the lactose metabolising system.[57] Although these mutants were isolated by several techniques (ultraviolet irradiation, nitrosoguanidine, acriflavine, or spontaneously) they all possessed the same enzymatic defects. The constitutive proteins Enzyme I and HPr were present but the lactose-specific components Enzyme II^lac, Factor III^lac and P-β-Gal were absent.

The significance of the simultaneous loss of the three lac-specific proteins and the effect of acriflavine (treatment of Lac$^+$ cells of C2 resulted in over 70 % of the survivors being Lac$^-$) as well as the spontaneous occurrence of Lac$^-$ variants was not realised when these observations were initially made.

PEP is required for the transport of lactose into group N streptococci by the PEP–PTS. Collins and Thomas,[7] Thomas,[81] and Thompson[83] presented evidence for the key role that pyruvate kinase plays in regulating the intracellular concentration of PEP and, thus, indirectly controlling the rate of lactose transport into the cells (Fig. 2). The regulatory function of the enzyme was suggested by the cooperative binding of its substrate PEP and allosteric activation by all of the glycolytic intermediates preceding 1,3-diphosphoglycerate.[81,82] Thus, with high levels of these intermediates, pyruvate kinase is activated and PEP is converted to pyruvate with generation of ATP. As the concentration of the activators decreases, however, the activity of pyruvate kinase also becomes less and the accumulating PEP can be used by a PTS. Therefore, as stated by Thompson,[83] the modulation of pyruvate kinase activity, through control of the intracellular PEP level, can determine the distribution of PEP into two separate pathways—one leading to ATP generation and the other leading to membrane-located PTSs.

Results obtained by Thompson and Thomas[85] and confirmed by Thompson[83] showed that starved cells of S. lactis ML$_3$ retained high levels of 3-PG, 2-PG, and PEP. Their observations were consistent with the demonstrated absence of the potential in vivo activators of pyruvate kinase.[82,85] Thompson[83] then very cleverly tested in vivo the pyruvate kinase activator hypothesis. He first blocked glycolysis by using p-chloromercuribenzoate (p-CMB) which selectively inhibited glyceraldehyde-3-phosphate dehydrogenase. This caused retention of those glycolytic metabolites preceding the blockage point (glucose-6-phosphate, fructose-6-phosphate, fructose-1,6-diphosphate, and triose phosphates). Since these intermediates are potential activators of pyruvate kinase, their presence leads to a depletion of 3-PG, 2-PG, and PEP, which occur after the point of inhibition. The inhibition by p-CMB was then reversed by adding dithiothreitol. This permitted conversion via glycolysis of the retained metabolites to 3-PG, 2-PG, and PEP. These PEP potential intermediates accumulated because there was a depletion in the pyruvate kinase activators, confirming that metabolic regulation of lactose utilisation in S. lactis ML$_3$ is tightly linked to the availability of PEP, which is controlled by pyruvate kinase.

Isolation of Constitutive Mutants for Lactose Utilisation

One of the principal functions of starters in the manufacture of cheese and cultured dairy products is the formation of lactic acid from lactose. Because of the economic importance of fermented products, the need to isolate new, fast acid-producing strains from milk or other natural sources has been suggested.[74] An alternative method would be the improvement of existing strains through genetic manipulation. One approach to obtain cells which have an enhanced ability to utilise lactose is to isolate lactose constitutive mutants. There is evidence that the addition of β-Gal to milk increases the rate of acid production by a starter culture.[17,69] Gilliland et al.[17] suggested that the lactose-utilising enzymes in lactic streptococci were not fully induced when the strains were grown in milk, thus maximal acid production by the cells did not occur, due to inefficient hydrolysis of lactose. Constitutive mutants, however, would have lactose-utilising enzymes that are functioning at full capacity.

By growing S. lactis C2 in a chemostat with lactose as the limiting growth factor, Schifsky and McKay[75] isolated variants believed to be lactose constitutive mutants. The P-β-Gal activity per unit mass of cells increased and reached a plateau after extended continuous growth. According to Horiuchi et al.[26] and Novick and Horiuchi[65] the increased enzyme activity during continuous culture is due to selection of lactose constitutive mutants. In E. coli, such mutants have been isolated and are capable of producing β-Gal to greater than 20 % of their cell protein. The mutants isolated from S. lactis C2 synthesised 1·5 to 4 times the amount of enzyme of the parent strain, but remained sensitive to repression by glucose. To obtain further evidence for constitutive lactose metabolism in these mutants, they were tested for growth in a medium containing lactobionic acid as the sole carbohydrate source.[75] Lactobionic acid does not induce the lac operon of E. coli, and is hydrolysed with difficulty by β-Gal.[40] According to Langridge,[40] mutants which grow on lactobionic acid are constitutive since they overcome the absence of substrate induction. Two mutations are required, one causing inactivation of the lac repressor and a second leading to an increase in the efficiency of the enzyme. A mutant of S. lactis C2 was able to grow on lactobionic acid, whereas the inducible parent culture was not, confirming the constitutive nature of the mutant. It was not determined whether the mutant had an increased efficiency of P-β-Gal activity or a mutation in a regulator (i gene) or an operator (o gene). These two regulatory mechanisms can now be analysed due to the recent development of genetic systems in lactic streptococci. The constitutive mutant of C2 initiated growth faster in milk than the parent culture,[75] and such strains may prove valuable in dairy fermentations as well as in the understanding of the genetic regulation of lactose metabolism in dairy streptococci.

Lactose Metabolism in Streptococcus thermophilus

Information is limited on the mechanisms and control of lactose metabolism in *Streptococcus thermophilus*. In a recent report, Somkuti and Steinburg[80] found that strains of *S. thermophilus* could be divided into three groups depending on their growth response to lactose, glucose, or galactose. Some strains fermented all three sugars, others utilised lactose and glucose only, and one strain grew on lactose alone. Reddy *et al.*[71] examined the lactose transport system in *S. thermophilus* and found that only galactose-adapted cells were induced to transport lactose. Since galactose is not utilised by all strains, differences in lactose transport systems and/or mechanisms of utilisation probably exist. Somkuti and Steinberg[79] screened 32 strains of *S. thermophilus* for lactose hydrolysing enzymes, and found β-Gal in 28 strains, both β-Gal and P-β-Gal in three strains, and only P-β-Gal in one strain. These enzymes appeared to be regulated by an induction–repression mechanism. Therefore, in *S. thermophilus* lactose is probably metabolised by a system similar to that reported for *E. coli* and/or a lactose PEP-dependent PTS system. Due to the extensive use of *S. thermophilus* in preparation of yoghurt and certain varieties of cheese, it is surprising that more is not known about the control of lactose metabolism in this organism.

EVOLUTIONARY ASPECTS

The natural habitat of the group N streptococci is considered to be green plant material even though an infallible source for these organisms is raw milk.[74] Hirsch[24] hypothesised that these organisms are of recent origin. He reasoned that although milk is not considered to be a normal habitat, they are evolving to it as an ecological niche. His reasoning included the saprophytic nature of the organisms, their lactose-fermenting ability, their habitat, and their antibiotic-producing ability. He postulated that *S. lactis*, as it is known to the dairy microbiologist, is a recently evolved, highly specialised member of a much wider complex of streptococci. Mundt[64] concurred and felt that in nature (plant material) lactose fermentation is not of paramount value. Streptococci which are unable to ferment lactose have been isolated from plant material by workers in the United Kingdom and the United States. More recently, Farrow and Garvie[15] suggested that cheesemakers have unknowingly selected those strains containing a highly efficient system for lactose utilisation and that these may be the only ones capable of metabolising lactose in the cheese vat at a rate fast enough for cheesemaking. If *S. lactis* is a highly specialised member of a much wider complex of streptococci, what kind of evolutionary change was required for the acquisition of a new metabolic capability of efficient lactose fermentation? To answer this question one must first examine the ease by which efficient strains lose the ability to ferment lactose.

LOSS OF LACTOSE-FERMENTING ABILITY

Variation in lactose metabolism by lactic streptococci was first observed in the late 1930s. Variants of *S. lactis* that did not ferment lactose were isolated from milk by Yawger and Sherman.[88] Okulitch and Eagles[68] found that successive transferring of *S. cremoris* 142 in a carbohydrate medium other than lactose or galactose resulted in the appearance of variants unable to ferment lactose. Hunter[27] also isolated variants of *S. cremoris* which failed to ferment lactose and were defective in galactose metabolism. In a 1951 paper, Hirsch[25] noted that *S. lactis* 354 became Lac⁻ after repeated subculturing in glucose broth. All attempts by various workers to define the precise conditions leading to the change from lactose fermenting (Lac⁺) to non-lactose fermenting (Lac⁻) were unsuccessful, even though many attempts resulted in the production of 100 % of the Lac⁻ cells. Thus, it was concluded that lactose-fermenting ability in some lactic streptococci is unstable under certain conditions of cultivation, i.e. growing the cells in the absence of lactose or galactose. The mechanism of loss of lactose metabolism was not known.

Lac⁻ mutants are defined as those isolates unable to ferment lactose. Since lactic streptococci are extremely fastidious in their nutritional requirements and do not grow well in defined or semi-defined medium, a complex medium consisting of tryptone, yeast extract, gelatin, sodium acetate, sodium chloride, lactose, bromocresol purple as pH indicator, and agar is used to screen for Lac⁻ derivatives.[55] On this medium Lac⁺ colonies are yellow due to acid production and Lac⁻ colonies are white. If lactose is omitted from the medium, the Lac⁺ cells will form white colonies or a lawn of white growth if a smear of Lac⁺ cells is spread over the surface of the plate. Therefore, when the medium contains lactose it is very easy to differentiate Lac⁻ from Lac⁺ colonies. When Lac⁻ variants are inoculated into a semi-synthetic medium containing lactose as the sole fermentable carbohydrate, they are unable to grow.[8] Implications have been made that some variants are truly Lac⁻ while others are not. The confusion may have resulted from the observations that Lac⁻ variants (at least from *S. lactis* C2) grew in milk to 2×10^8 colony forming units per ml though no acid was produced.[50] These same mutants are clearly Lac⁻, however, based on the white colony phenotype on lactose indicator agar and on their inability to utilise lactose as a sole carbohydrate source in a semi-synthetic medium.

The high incidence of spontaneous loss of lactose metabolism suggested that lactic streptococci could be carrying an extrachromosomal genetic element responsible for the cells' ability to ferment lactose. The loss of this element, termed a plasmid, would cause the cell to become Lac⁻. If this were the case, then lactic streptococci would be expected to form plasmid-negative variants as a result of occasional errors in replication.[66] This spontaneous loss of lactose metabolism by lactic streptococci was noted by early workers and confirmed by McKay *et al.*[55] The frequency of such variants can often be increased by treatment with certain chemical agents such as acridine dyes which selectively prevent plasmid replication.[23] The appearance of Lac⁻ mutants increased with acriflavine treatment in *S. lactis*, *S. cremoris*, and *S. diacetylactis*.[55] In *S. lactis* C2, it was shown that acriflavine appeared to competitively favour the growth of spontaneously occurring Lac⁻ cells and to act directly in the conversion of Lac⁺ to Lac⁻ cells.

The high incidence of the spontaneous loss of lactose metabolism coupled with the effect of acriflavine is presumptive evidence for the involvement of plasmid deoxyribonucleic acid (DNA) in lactose fermentation by lactic streptococci. Furthermore, one may hypothesise that the highly specialised strains of lactic streptococci evolved from the much wider complex of streptococci by acquiring lactose-fermenting ability via transfer of genetic material from another species. There is a precedent for this hypothesis. A plasmid isolated from *Salmonella tennessee* carries determinants for the fermentation of lactose[29] as does a plasmid isolated from a *Proteus* strain.[87] Both of these genera do not normally ferment lactose, thus these variants are presumed to be instances where genes were acquired from unrelated sources.[73]

SIMULTANEOUS LOSS OF LACTOSE-FERMENTING ABILITY AND PROTEINASE ENZYME ACTIVITY

When strains of lactic streptococci lose their ability to ferment lactose, they simultaneously lose certain other metabolic functions. Galactose metabolism becomes defective due to loss of ability to utilise the sugar via the lactose PEP–PTS. Proteinase activity is also lost. This activity is required for optimal growth and acid production when streptococci are grown in milk, since this system provides needed nitrogenous compounds from casein.

While it was previously known that a strain could lose its proteinase activity and still retain its lactose utilising system (Lac$^+$ Prt$^-$) not until work by Molskness *et al.*[60] and McKay and Baldwin[50] using Lac$^-$ mutants of *S. lactis* C2 was it known that Lac$^-$ cells were also Prt$^-$. The latter workers observed that the parent culture when grown in milk at 32 °C reached a population of about $3\cdot0 \times 10^9$ colony-forming units per ml in 18 h. The pH decreased from 6·8 to 4·6 and the organism exhibited proteinase activity as measured by tyrosine released from casein. The maximal count of the Lac$^-$ Prt$^-$ mutant was about $2\cdot0 \times 10^8$ colony-forming units per ml. Although the organism grew in milk, no proteolysis was detected after 25 h and the pH decreased to 6·4. The mutant was then grown in milk supplemented with glucose or nitrogenous compounds. In glucose-supplemented milk, the mutant remained defective in proteolytic enzyme activity but did lower the pH to 4·5 in 25 h, indicating that glucose was being fermented. When the milk was supplemented with exogenous nitrogen compounds, the mutant still did not ferment lactose as evidenced by no major decrease in pH. If the milk was supplemented with both glucose and nitrogenous compounds to bypass both enzymatic defects, rapid acid production was restored. Lac$^-$ derivatives from *S. diacetylactis* 18-16 were also Prt$^-$, but Lac$^-$ variants from *S. cremoris* B1 were Prt$^+$. Lac$^-$ Prt$^-$, Lac$^-$ Prt$^+$, or Lac$^+$ Prt$^-$ derivatives can occur from Lac$^+$ Prt$^+$ strains of lactic streptococci. Plasmid participation provides an explanation for spontaneous loss of the *lac* and *prt* genes.

TRANSDUCTION OF LACTOSE-FERMENTING ABILITY

Bacterial transduction is observed when a temperate bacteriophage transfers a portion of the host's bacterial chromosome to a recipient bacterium, where it may become incorporated into the chromosome. A temperate bacteriophage was demonstrated in *S. lactis* C2 by exposing the cells to ultraviolet (UV) irradiation, inoculating the irradiated cells into broth, and following the change in turbidity. The turbidity increased for about 2 h at which time the suspension lysed, as indicated by rapid clearing of the culture, Although an indicator strain was not isolated, the existence of phage was confirmed by obtaining electron micrographs of the phage particles.[37,49] In examining the temperate phage induced from *S. lactis* C2, it was found that two phage particles were induced which differed in head size.[37] The two phages had head diameters of 60 nm and 70 nm. The amount of DNA that could be packed in these phage heads was calculated to be about 22.6×10^6 daltons and 23.8×10^6 daltons. This phage was used in developing a genetic transfer system in group N streptococci.

Further results indicated that a potential transducing phage had been isolated.[56] UV induced lysates prepared from S. *lactis* C2 converted lactose-, maltose-, or mannose-negative recipient cells of this strain to the respective carbohydrate-positive phenotype. Cell-to-cell contact was not required, ruling out conjugation as the mechanism of genetic transfer, and treatment of the lysate with nuclease had no effect on cell conversion from Lac⁻ to Lac⁺, indicating that the genetic change was not mediated by free DNA (transformation). Transduction was further supported by the observation that the number of Lac⁺ transductants obtained was proportional to the amount of phage lysate used in transduction trials. The demonstration of a genetic transfer system in S. *lactis* C2 provided a tool for examining the spontaneous loss of lactose metabolism and other traits in this organism.

The number of Lac⁺ transductants observed using a lysate prepared from S. *lactis* C2 was low (50 to 100 per ml). However, UV induction of these transductants yielded lysates giving a higher frequency of transfer of lactose-fermenting ability (75 000 per ml). The transfer of maltose- or mannose-fermenting ability did not exhibit this phenomenon. There are at least two different mechanisms to explain high frequency transduction (HFT) of lactose-fermenting ability. Analogous to E. *coli* lambda Gal⁺ transductants,[62,63] the prophage could be integrated in the chromosome adjacent to the *lac* genes; however, the unstable nature of lactose metabolism in S. *lactis* C2 suggested the trait is plasmid-linked. An alternative possibility is that the HFT lysates were being formed by recombination between phage and plasmid DNA.[66]

Lac⁻ mutants of S. *lactis* C2 are also Prt⁻. When these Lac⁻ Prt⁻ mutants were transduced with the temperate phage from the Lac⁺ Prt⁺ parent culture, approximately equal numbers of Lac⁺ Prt⁺ and Lac⁺ Prt⁻ transductants were obtained.[50] The latter produced acid slowly in milk due to a deficiency in the proteinase system, whereas the former resembled the parent culture in acid-producing ability and proteinase activity. Similar observations were made by Molskness et al.[60] They found that Lac⁺ transductants were similar to the wild type with respect to lactose fermentation and level of P-β-Gal, but all their transductants were Lac⁺ Prt⁻.

LACTOSE METABOLISM AND LINKAGE TO PLASMID DNA

Plasmids are defined as autonomous, replicating, extrachromosomal, genetic elements. Plasmids are unstable entities within the bacterial cell, separate from the chromosome, which when lost cause the cell to lose the functions dictated by the plasmid DNA (Fig. 3(A)). Plasmid participation provides a mechanism for explaining the spontaneous loss of the *lac* and *prt* genes or *prt* genes alone in *S. lactis* C2, as well as the appearance of Lac$^+$ Prt$^+$ or Lac$^+$ Prt$^-$ transductants. The simplest model postulates that *S. lactis* C2 has at least three types of plasmids—Lac, Prt, and Lac Prt—which could explain the behaviour of spontaneous variants and transductants.

The isolation of plasmid DNA from lactic streptococci involves separating plasmid DNA from chromosomal DNA as described by Cords et al.,[9] and modified by Klaenhammer et al.[38] DNA isolated from the cells was mixed with caesium chloride and ethidium bromide and subjected to ultracentrifugation. Ethidium bromide inserts between the base pairs of the DNA molecules decreasing the density of the DNA. Since plasmid DNA occurs as a tightly coiled covalently closed circle, the amount of ethidium bromide that can be inserted is limited in contrast to other species of DNA that may be present. Thus, the plasmid DNA can be separated from the other forms of DNA as a denser band following density gradient centrifugation. When DNA has been labelled with [^{3}H]thymidine the

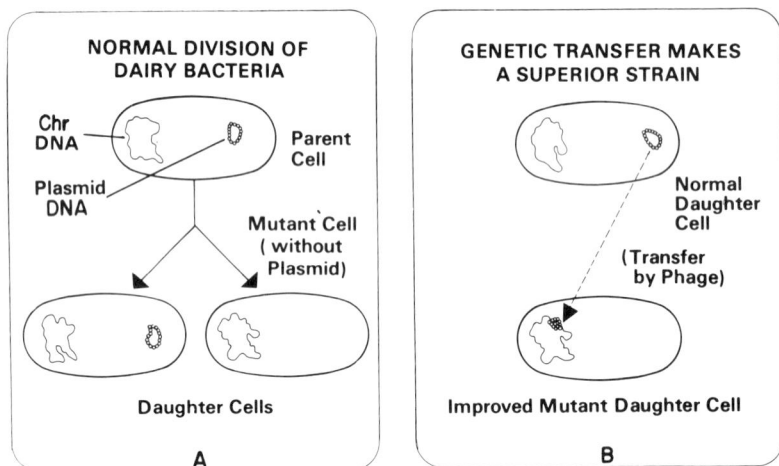

FIG. 3. Plasmid loss from a bacterial cell (A) and stabilisation of a metabolic trait linked to plasmid DNA by integration of the plasmid or a portion of it into the bacterial chromosome (B).

distribution of the label in the fractions can be determined. Typical gradient profiles from *S. lactis* C2 (Lac$^+$ Prt$^+$) and Lac$^-$ Prt$^-$ or Lac$^+$ Prt$^-$ derivatives are shown in Fig. 4. The existence of plasmid DNA in these strains are indicated by a satellite band (left peak) separate from the chromosome band (right peak). As plasmid isolation procedures improved, dye-buoyant density gradient centrifugation of the cleared lysates resulted in a distinct satellite DNA band visible in the gradient upon UV light illumination. The plasmids from a particular strain could thus be easily isolated without [^{3}H]thymidine labelling.

The different size classes of plasmids present in these satellite bands were first determined using electron microscopy.[9] Ethidium bromide was

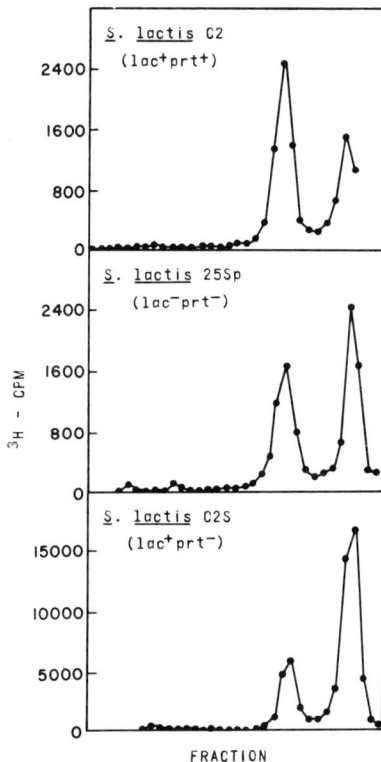

FIG. 4.   Elution profile of caesium chloride–ethidium bromide density gradients of DNA isolated from. Lac$^+$ Prt$^+$ *S. lactis* C2 and its Lac$^-$ Prt$^-$ or Lac$^+$ Prt$^-$ derivatives. A satellite band representing plasmid DNA appears as the left peak in each profile, separate from the right peak of chromosomal DNA.[47]

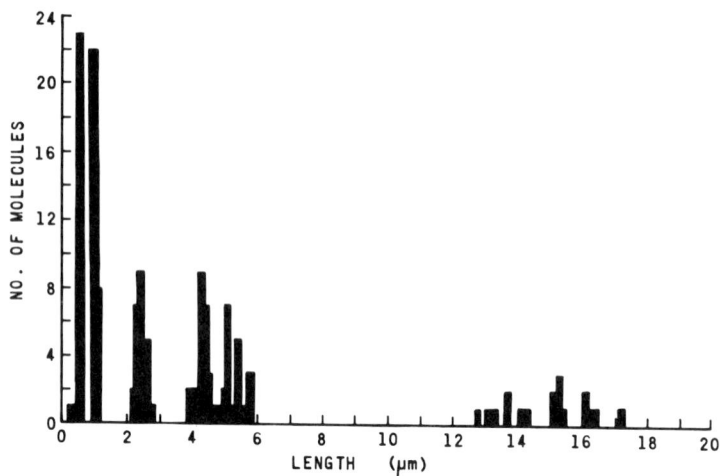

FIG. 5.   Distribution of contour lengths of circular molecules of DNA from Lac$^+$ Prt$^+$ S. *lactis* C2. Five different plasmid sizes were observed, corresponding to molecular weights of 1×10^6, 2×10^6, 5×10^6, 10×10^6, and 30×10^6. The molecular weights are determined by multiplying the observed contour length by 2.07×10^6 daltons per μm.[51]

removed from the plasmid sample, caesium chloride dialysed out, and the DNA sample was aged. During this process the tightly coiled plasmid molecules became nicked in one of the strands of the DNA, resulting in an open circle of DNA. If both strands become nicked, the molecule becomes linear. Electron micrographs of the open circular forms of the DNA were obtained and contour lengths of the molecules determined. A histogram of length versus number of molecules measured was prepared to visualise the distribution of plasmid sizes within a strain. Figure 5 is a histogram showing the distribution of plasmid sizes found in S. *lactis* C2. Five different classes of plasmid sizes were observed having molecular weights of 1×10^6, 2×10^6, 5×10^6, 10×10^6 and 30×10^6 daltons. As will be shown later by agarose gel electrophoresis, the 10×10^6 dalton plasmid band actually was composed of two different sized plasmids.

Mutants were examined for plasmid loss concomitant with loss of ability to ferment lactose or produce the proteinase enzymes. In S. *lactis* C2, the 10×10^6 dalton plasmid band was missing when the strain became Lac$^+$ Prt$^-$, suggesting it is linked to proteinase activity in this strain.[51] The size of plasmids observed in Lac$^+$ Prt$^+$ S. *cremoris* HP were 1.8×10^6, 3.3×10^6, 8.1×10^6, 18.7×10^6, and 29.7×10^6 daltons.[41] Lac$^+$ Prt$^-$ variants of HP missing the cell-wall bound proteinase[14] had lost the 8·1 megadalton (Mdal) plasmid.[41]

Based on the Lac⁻ Prt⁻ phenotype, one would expect to see two plasmids missing if individual plasmids were responsible for the cells' ability to ferment lactose and produce proteinase; this was not the case. When *S. lactis* C2 became Lac⁻ Prt⁻ the 30 Mdal plasmid was missing, yet the variants still retained what are believed to be the proteinase plasmids. This was best illustrated by agarose gel electrophoretic patterns of plasmid DNA isolated from Lac⁺ Prt⁺ *S. lactis* C2 and Lac⁺ Prt⁻ or Lac⁻ Prt⁻ variants from this strain.[38] *S. lactis* C2 was shown to contain six resident plasmids. The plasmid band corresponding to a 10 Mdal observed by electron microscopy was resolved into two plasmids having molecular weights of $12·5 \times 10^6$ and 18×10^6. The Lac⁺ Prt⁻ variant was missing the latter two plasmids but only the 30 Mdal plasmid was absent in the Lac⁻ Prt⁻ strains. To determine a relationship between lactose-fermenting ability and the 30 Mdal plasmid, a Lac⁻ Prt⁻ mutant of *S. lactis* C2 was cured of the prophage(s) as well as the six resident plasmids. Since this mutant, designated LM0230, was devoid of plasmid DNA it functioned easily as a recipient in transduction experiments and was used to determine which, if any, of the plasmids were acquired by Lac⁺ Prt⁺ and Lac⁺ Prt⁻ transductants. In both phenotypes a single plasmid having a molecular weight of about 23×10^6 was transferred into the recipient. This size plasmid molecule was unique to the Lac⁺ Prt⁺ or Lac⁺ Prt⁻ transductants since it was not observed in the parent culture. When the Lac⁺ transductants of LM0230 became Lac⁻ (spontaneous or acriflavine treatment) the 23 Mdal plasmid was no longer observed.[38,52] Thus, one of the criteria necessary to link a particular metabolic function to plasmid DNA was met.

Since the transducing phage could only accommodate a piece of DNA having a molecular weight of about 23×10^6, it was evident that the transduced lactose plasmid had to be smaller than the Lac plasmid present in the parent culture. Plasmid transduction has been described in a number of microorganisms.[2,6,32] Shipley and Olsen[77] described transduction with a phage that was too small to accommodate the entire plasmid genome, causing the formation of a smaller plasmid having only a portion of the genetic complement of the original parent plasmid. They termed the process transductional shortening and assumed that the process involved fragmentation of the plasmid DNA during packaging into the phage head followed by recirculation in the recipient to form a new autonomous replicating plasmid. A similar phenomenon of transductional shortening appears to operate in *S. lactis* C2. It was found that the Lac plasmid from C2 carried genetic markers for resistance to arsenate, arsenite, chromate and for sensitivity to copper.[11] The copper sensitivity marker appeared to be on that portion of the plasmid which was lost during the transductional shortening process.

Plasmid-linked lactose-fermenting ability in C2 was further sub-stantiated.[11] According to Arber,[3] chromosomal genes in a transducing lysate are less sensitive to UV inactivation than are plasmid-linked genes. The transducing frequency for the chromosomal genes may be stimulated by exposure of the transducing lysate to low UV doses, while plasmid-linked determinants will show an exponential decrease in transducing ability. Transductional analysis was used to confirm that lactose fermentation in C2 was plasmid mediated. The transducing phage was exposed to a series of UV doses. Samples were removed after each UV dose and assayed for ability to transduce lactose- and maltose-fermenting ability. The results indicated that small doses of UV irradiation increased the frequency of transduction of maltose-fermenting ability while decreasing, in an exponential fashion, the transduction of lactose-fermenting ability. This response would be expected of chromosomal and plasmid mediated traits, respectively.

The Lac plasmid from C2, as well as from other lactic streptococci appears to be more difficult to isolate than are other plasmids present in these strains. The scarcity of the Lac plasmid molecules present in earlier preparations[52] was found by Klaenhammer *et al.*[38] to be due to the 1·5 h lysozyme treatment required to obtain lysis of the cells. Further study with a modified Elliker broth[13] determined that a lysozyme digestion period beyond 20 mins resulted in the loss of the Lac plasmid species, but not of the other resident plasmids. This suggested that Lac plasmid loss during isolation procedures was caused by nuclease activity. To account for the specificity of nuclease activity on Lac plasmid DNA, Klaenhammer *et al.*[38] proposed that it could be due to increased availability of general binding sites for endogenous nuclease on the Lac plasmid or alternatively, endogenous nuclease binding may be site specific for nucleotide sequences present on the Lac plasmid. Involvement of a site specific endonuclease could account for the loss of only Lac plasmid DNA. Further research is needed to clarify the labile nature of Lac plasmid DNA. The study of plasmids in *S. lactis* C2 resulted in the development of a more efficient lysis procedure for isolation of group N streptococcal plasmid DNA. This procedure, coupled with agarose gel electrophoresis, provided a rapid method for analysis of plasmid DNA composition in these organisms.

When strains of lactic streptococci other than *S. lactis* C2 were examined for plasmid linkage of lactose-fermenting ability, a correlation was again noted. *S. cremoris* B1 was examined by Anderson and McKay[1] and found to contain two plasmids having molecular weights of 36×10^6 and 9×10^6. Analysis of Lac⁻ variants of this strain suggested lactose metabolism was linked to the 36 Mdal plasmid; Lac⁻ strains were missing both or only the 36 Mdal plasmid. In contrast to other Lac⁻ variants, these were Lac⁻ Prt⁺. Proteinase activity did not appear to be plasmid-linked in *S. cremoris* B1;[55] possibly this is a strain in which a proteinase plasmid has been integrated into the chromosome thus stabilising the characteristic.

Using electron microscopy, Efstathiou and McKay[10] initially reported that in *S. lactis* strains C10, ML₃, and M18 lactose metabolism was determined by an 18 to 22 Mdal plasmid, and proteinase activity was determined by an 8 to 10 Mdal plasmid. The determination of the molecular weights of the plasmids in C10, ML₃ and M18 by agarose gel electrophoresis, however, revealed the absence of these plasmids in the parent strains.[38] Since Klaenhammer *et al.*[38] had shown that analysis by electron microscopy and agarose gel electrophoresis produced values which were in agreement for the molecular weights of plasmids from *S. lactis* C2 and *S. cremoris* B1, it appeared that the assignment of lactose metabolism and proteinase activity to the 18 to 22 Mdal and 8 to 10 Mdal plasmids, respectively, was in error. Subsequent work by Kuhl *et al.*[39] showed that the Lac⁻ Prt⁻ variants of C10, ML₃, and M18 were each missing a single plasmid having molecular weights of 40×10^6, 33×10^6, and 45×10^6, respectively. A similar observation was noted in *S. lactis* C2, in which the predominant mutant phenotype isolated was a Lac⁻ Prt⁻ mutant missing a 30 Mdal plasmid. Transductional analysis supported the conclusion that both lactose metabolism and proteinase activity were linked to the 30 Mdal plasmid in C2. Yet, it has been shown that a Lac⁺ Prt⁻ mutant of C2 is missing a 12·5 Mdal and an 18 Mdal plasmid, suggesting that one or both of these plasmids are linked to proteinase activity. In Lac⁺ Prt⁺ *S. cremoris* HP proteinase activity appears to be linked to an 8·1 Mdal plasmid. In addition, there are other reports that proteinase activity may be plasmid-borne and independent of lactose-fermenting ability.[42] The nature of plasmid interrelationships between lactose metabolism and proteinase activity in lactic streptococci remains unresolved. In some strains, C2 for example, both a Prt and a Lac Prt plasmid appear to be present. It is not known why a strain can lose the Prt plasmid(s) and become Lac⁺ Prt⁻; yet, if the Lac Prt plasmid is lost, the strain becomes Lac⁻ Prt⁻ even though the Prt plasmid(s) is retained. This phenomenon is further complicated by the observation that Lac$^+$ Prt$^+$ transductants of C2 can occur in which a single plasmid has been added to a Lac$^-$ Prt$^-$ recipient. One could hypothesise a variety of regulatory interchanges occurring among the plasmid and chromosomal DNA. The lactic streptococci have not yet been examined for the presence of transposable genetic elements. Transposons may be involved in the instability of Prt. As more research is conducted in this area, the reason for the above observations should become more clear.

The linkage of plasmid DNA to the ability of lactic streptococci to ferment lactose has been implicated for other strains. LeBlanc et al.[43] presented evidence that S. lactis DR1251 was dependent upon a 32 Mdal plasmid for lactose utilisation. S. diacetylactis strain 18–16 becomes Lac$^-$ when a 41 Mdal plasmid is lost from the parent culture, and strain DRC1 becomes Lac^d Prt$^-$ upon losing a 31 Mdal plasmid.[34] When nine strains of S. cremoris (HP, AM$_2$, ML$_1$, WC, C$_3$, R$_1$, E$_8$, KH, and Wg$_2$) were examined for plasmid DNA, they were shown to possess a diversity of plasmid sizes.[41] In contrast to S. lactis and S. diacetylactis, few Lac$^-$ variants could be isolated from the S. cremoris strains and no definite conclusion was made on the relationship between plasmid DNA and the ability of these strains to ferment lactose. Exhaustive curing experiments with AM$_2$ using acridine dyes, elevated temperature, and chemostat growth did not yield Lac$^-$ variants. The inability to obtain Lac$^-$ variants under conditions known to facilitate plasmid elimination suggested that lactose metabolism is not plasmid mediated in AM$_2$. Although it has been found that S. cremoris strains are more difficult to investigate with respect to plasmid DNA, studies are needed in this area.

Linkage of lactose metabolism to plasmid DNA in certain strains of lactic streptococci has also been confirmed by the recent demonstration of a conjugal transfer system in this group of organisms. Gasson and Davies[16] briefly described the conjugal transfer of lac genes from a Lac$^+$ donor to a Lac$^-$ recipient. About the same time, Kempler and McKay[35] confirmed that the 41 Mdal plasmid missing in Lac$^-$ derivatives of S.diacetylactis 18-16 was linked to lactose metabolism. Lac$^+$ cells of 18-16 were shown to transfer the 41 Mdal plasmid to LM0230 (a Lac$^-$ plasmid-cured derivative of S. lactis C2) via a process requiring cell-to-cell contact. The acquisition of the 41 Mdal plasmid by LM0230 resulted in a Lac$^+$ phenotype and demonstrated the conjugal transfer of plasmid DNA in dairy streptococci. Using LM0230 as a recipient, the conjugal transfer of a lactose plasmid was subsequently shown for S. lactis ML$_3$[54] as well as for S. lactis C$_2$O and S. diacetylactis strains DRC3, 11007 and WM4.[53] The discovery of genetic transfer systems for the lactose plasmid (transduction and conjugation) coupled with the Arber experiment using phage induced from *S. lactis* C2, leaves no doubt that the spontaneous loss of lactose-fermenting ability observed in these organisms is due to plasmid DNA. The origin of the lactose plasmid in these strains is unknown; however, the acquisition of a Lac plasmid by these streptococci would certainly allow the emergence of highly specialised strains such as those selected for cheesemaking. Since Lac$^+$ streptococci and staphylococci appear to utilise lactose by the same biochemical reactions, it is tempting to speculate that there could be exchange of genetic material between the two groups. Whether this could account for the similarity in mechanisms of lactose utilisation is not known but this does present many interesting questions.

STABILISATION OF LACTOSE FERMENTATION

Knowledge that lactose-fermenting ability is linked to plasmid DNA suggested the possibility of stabilising this important metabolic function in the lactic streptococci. The integration of the lactose plasmid from these strains into their host chromosomes would accomplish this objective, (Fig. 3(B)). McKay and Baldwin[48] were able to isolate transductants from *S. lactis* C2 which were stabilised for lactose metabolism. This stabilisation was confirmed by several observations. First, it was found that acriflavine treatment was ineffective in causing the conversion from Lac$^+$ to Lac$^-$ in these particular variants. The parent culture possessed about 88 % Lac$^-$ cells after six consecutive transfers in the presence of the curing agent while normal transductants (those acquiring a demonstrable lactose plasmid) contained from 11 to 23 % Lac$^-$ cells after 10 successive transfers. The second observation was that no Lac$^-$ variants could be found when one of the stabilised strains was grown in a chemostat for 165 h, whereas the parent culture contained over 90 % Lac$^-$ cells. McDonald[46] also observed the latter phenomenon for the continuous growth of other lactic streptococci and concluded that before continuous culture techniques could be used for starter culture production it would be necessary to select strains that would not give rise to undesirable variants. The stabilised Lac$^+$ variants may be such strains, at least for maintenance of lactose-fermenting ability.

The stability of lactose metabolism could have been due to a mutation resulting in extreme stability even under conditions that were non-selective for maintenance of the plasmid.[45] This was ruled out by showing that the lactose plasmid normally present in Lac$^+$ transductants, was missing in the stabilised strains when examined by agarose gel electrophoresis. Since acridine dyes are capable of eliminating plasmids in the autonomous state but have no effect on plasmids integrated into the chromosome,[23,28] the stability of the Lac⁺ phenotype coupled with the absence of visible lactose plasmid suggested that the plasmid or a portion of it had been integrated. This was confirmed by performing an Arber experiment. The genes responsible for lactose metabolism in the stabilised strain selected for further testing behaved as chromosomal-linked material instead of as plasmid-linked determinants.[48] Thus, lactose metabolism normally unstable in *S. lactis* C2 due to its plasmid linkage had become stabilised by incorporating the *lac* genes into the chromosome. Since this phenomenon occurred in *S. lactis* C2, it would not be surprising to find its occurrence in other strains of lactic streptococci. As more strains are examined it is likely that some will be found in which lactose metabolism is not linked to plasmid DNA, because the *lac* genes have been integrated into the chromosome. As described earlier, *S. cremoris* AM_2 may represent such a strain.

If lactic streptococci are of recent origin and are evolving to milk as their natural habitat as proposed by Hirsh[24] it is logical to assume that this adaptation is due to plasmid DNA. Thus plasmid DNA may be considered to be a vehicle in evolution, in that it is the intermediate form between strains unable to utilise lactose and those with chromosomally linked *lac* genes. As these 'specialised' strains continue to evolve in milk, it is also logical to assume that indispensable traits, such as efficient lactose fermentation, would eventually become stabilised as chromosomal genes. Of the four stabilised Lac⁺ variants isolated from *S. lactis* C2, three were Lac⁺ Prt⁻ and one was Lac⁺ Prt⁺; however, the proteinase activity observed in the latter strain was about half the value expressed by the parent culture. Even though chromosomal integration of the lactose plasmid, or a portion of it, had occurred, Cheddar cheese made with the Lac⁺ Prt⁺ variant was equivalent, if not better in quality than cheese made using *S. lactis* C2.[33]

PARTIAL LACTOSE-FERMENTING REVERTANTS AND MODELS FOR LACTOSE UTILISATION

Various cultural conditions are known to result in the loss of lactose-fermenting ability in group N streptococci and this phenomenon has been shown to occur through the loss of plasmid DNA from the cell. Partial lactose-fermenting revertants can be detected when a lawn of Lac⁻ cells is incubated on the surface of lactose indicator agar. These revertants are considered partial because they lack the parental Lac$^+$ phenotype. The appearance of the revertants is a pale yellow colony surrounded by a minute yellow zone in contrast to a bright yellow colony and large yellow zone produced by Lac$^+$ parent cultures.

These revertants were studied to further understand the mechanisms and control of lactose metabolism in this group of microorganisms. Previous workers[24,27,43,68,88] considered the change from Lac$^+$ to Lac$^-$ in *S. lactis* and *S. cremoris* to be irreversible. Procedures described by Cords and McKay[8] resulted in the detection of reversion from Lac$^-$ to a partial lactose-fermenting phenotype in most strains examined. The partial Lac$^+$ revertants grew as cryptic mutants on a complex medium containing lactose as the primary fermentable carbohydrate. In other words, the cells grew as if the entry rate of lactose into the cell was the limiting step for growth. It was subsequently found that the partial revertants were unable to concentrate [^{14}C]TMG and that the uptake and hydrolysis of ONPG were concentration dependent. A transport defect was confirmed by complementation tests in which the partial revertants were shown to be defective in Enzyme II^{lac} and Factor III^{lac}. The cryptic nature of the revertants was further substantiated by showing that whole cells exhibited negligible P-β-Gal activity, yet significant activity was noted in toluene-treated cells, indicating impairment of the permeability barrier. The partial Lac$^+$ revertants thus appear to possess P-β-Gal but lack Enzyme II^{lac} and Factor III^{lac} of the PEP–PTS for lactose utilisation. The P-β-Gal was induced only when the cells were grown in the presence of lactose; galactose did not serve as an inducer. This is in contrast to the Lac$^+$ parent cultures in which galactose, as well as lactose, will induce P-β-Gal activity.

A model was proposed to account for the partial Lac$^+$ revertants obtained from Lac$^-$ derivatives of *S. lactis* C2 and *S. cremoris* B1.[1,8] The location of the genetic determinants for the Lac-specific enzymes (Enzyme II^{lac}, Factor III^{lac}, and P-β-Gal) of the PTS in the unstable Lac$^+$ strains is on plasmid DNA. The site of galactose-6-phosphate induction is located on the same plasmid. Loss of this plasmid results in the inability to metabolise lactose via the PTS, with a phenotypic change from Lac$^+$ to Lac$^-$. A Lac$^-$ strain may then undergo a chromosomal alteration(s) of unknown nature, resulting in the ability to slowly metabolise lactose. The partial lactose utilising system requires lactose transport and phosphorylation as well as P-β-Gal activity. Lactose, or a lactose derivative, is the inducer of the chromosomal P-β-Gal gene. The rate of partial lactose metabolism is limited by lactose transport and/or the phosphorylation process.[1] This model is supported by the following observations. There are instances of a second set of β-Gal genes in *E. coli* which appear after Lac⁻ derivatives are exposed to lactose.[86] Reeve and Braithwaite[72] have shown that *Klebsiella aerogenes* carries two sets of *lac* genes (one plasmid associated and the other chromosomal-linked). The Lac plasmid was responsible for the strong lactose-positive phenotype found in many of the strains. Two distinct systems of lactose metabolism may also exist in many strains of lactic streptococci.

   S. cremoris B1 contains two plasmids of molecular weight 9×10^6 and 36×10^6. Lac⁻ variants were isolated missing the 36×10^6 dal or both plasmids, suggesting the 36×10^6 dalton plasmid is linked to lactose-fermenting ability. Partial lactose-fermenting revertants having properties similar to those previously described occurred from both mutant types. In addition, a third Lac⁻ variant (DA1) was isolated which contained both the 9 and 36 Mdal plasmids. This mutant gave rise to partial revertants, but full revertants possessing the parental Lac⁺ phenotype also occurred. The basic difference between this Lac⁻ strain and the other Lac⁻ derivatives which do not give rise to full revertants is the retention of the 36 Mdal plasmid linked to lactose metabolism in B1. The phenotypic difference between Lac⁺ B1 and Lac⁻ DA1 was explained by a single deficiency in a lac-specific enzyme of the PTS. DA1 was defective in Enzyme II^{lac} and thus was unable to catalyse the transport and phosphorylation of lactose. The Lac⁻ variants lacking the 36×10^6 dalton plasmid were deficient in Factor III^{lac}, Enzyme II^{lac}, and P-β-Gal. The ability of DA1 to revert to a full lactose metabolising phenotype resulted in reactiviation of Enzyme II^{lac}. The presence of the three lac-specific enzymes in full revertants of DA1 that possessed the 36 Mdal plasmid, in conjunction with the comparative absence of these enzymes in the Lac⁻ and partial revertants not possessing this plasmid, strongly suggest that the genetic determinants for all three lac-specific enzymes are located on the 36 Mdal plasmid. In addition, the observation that all three independently isolated Lac⁻ variants from B1 revert to the partial Lac⁺ phenotype is evidence that the gene(s) enabling slow lactose metabolism is located on the only common genetic determinant, the host chromosome.

   Growth on galactose is more effective in inducing P-β-Gal than is lactose in wild type Lac⁺ cells, but in the partial Lac⁺ revertants lactose was the inducer, not galactose. This suggested that two separate P-β-Gal genes exist. A plasmid locus was suggested for one P-β-Gal gene and chromosomal locus was postulated for a second P-β-Gal gene. The latter would be required to allow a Lac⁻ mutant without a lactose plasmid to acquire a partial Lac⁺ phenotype and concomitant P-β-Gal activity.

When grown on lactose, the partial revertant from DA1 exhibited about 5·6 times the P-β-Gal activity of the parent culture.[1] It was suggested that both the chromosomal and plasmid P-β-Gal genes were being expressed in this partial revertant. Since galactose did not elicit this response, it appeared that lactose was transported and/or phosphorylated by a system other than the wild type lactose-PTS. Intracellular lactose, or a derivative, would induce the chromosomal P-β-Gal gene. The chromosomal P-β-Gal then cleaves lactose phosphate within the cell to glucose and galactose-6-phosphate. This internally formed galactose-6-phosphate would be capable of inducing the plasmid associated P-β-Gal. This could also explain why lactose and not galactose induces the chromosomal P-β-Gal in partial revertants lacking the lactose plasmid. Both sugars would be prevented from forming their phosphorylated derivatives via the lactose-PTS, but lactose phosphate could be formed by the alternate system and upon hydrolysis yield galactose-6-phosphate which could serve as the inducer. In other words, lactose appears to be the inducer only because partial revertants can obtain galactose-6-phosphate from it; they cannot form this compound from galactose. Galactose is metabolised in partial Lac$^+$ revertants via the Leloir pathway, but not the lactose-PTS.

Earlier in this chapter it was mentioned that S. *lactis* is a highly specialised member of a much wider complex of streptococci which cannot, or can only weakly, ferment lactose. Efficient utilisation of lactose appears to be due to acquisition of a plasmid possessing the genetic information for a lactose-PTS. As strains acquire the Lac plasmid it is possible that the gene(s) responsible for the weak ability to ferment lactose became silent and unexpressed. When an efficient strain loses the lactose plasmid and becomes Lac$^-$, exposure of the cells to lactose could select for mutations allowing expression of the dormant gene(s) and the resultant partial Lac$^+$ revertants. If the above rationale is accepted, one may further postulate that strains may exist possessing both systems of lactose utilisation. Such strains, upon losing the lactose plasmid, would immediately exhibit the partial Lac$^+$ phenotype in all the cells. In this regard, it was reported by Kempler and McKay[34] that when S. *diacetylactis* DRC1 lost a 31 Mdal plasmid, the strain was still able to weakly ferment lactose (Lac^d). Farrow and Garvie[15] have shown that a wild strain of S. *lactis* possessed both β-Gal and P-β-Gal activity. A recent survey by Okamoto and Morichi[67] for the distribution of the two enzymes among 40 lactic streptococci confirmed P-β-Gal to be the predominant enzyme and β-Gal activity to be very weak or not detected in many of the strains, while others were found to possess significant levels of both enzymes. These observations support the conclusion that more than one mechanism of lactose utilisation can simultaneously exist in group N streptococci. *S. lactis* 7962, although considered an atypical *S. lactis*, possesses only β-Gal.[15,58,59] Okamoto and Morichi[67] suggested that a strain having β-Gal and P-β-Gal may be an intermediate type between typical lactic streptococci possessing only P-β-Gal and strains like *S. lactis* 7962 which contain only β-Gal. In the oral streptococci (*S. salivarius*, *S. sanguis*, and *S. mutans*) both systems have also been shown to simultaneously exist and to be coinduced by lactose or galactose.[18] Thus several mechanisms and/or combinations of mechanisms which function separately or together, appear to be responsible for the ability of *S. lactis*, *S. cremoris*, and *S. diacetylactis* to ferment lactose.

The possession of the PEP–PTS for lactose utilisation appears to be a prerequisite for rapid homolactic fermentation of lactose by group N streptococci.[42] The state of our knowledge concerning the regulation of lactose metabolism in these organisms is in its infancy and as further research is conducted, the concepts involved should become more clear.

REFERENCES

1. ANDERSON, D. G., and McKAY, L. L., *J. Bacteriol.*, 1977, **129**, 367.
2. ANDERSON, E. S., and NATKIN, E., *Mol. Gen. Genet.*, 1972, **114**, 261.
3. ARBER, W., *Virology*, 1960, **11**, 273.
4. BECKWITH, J. R., *Science*, 1967, **156**, 597.
5. CITTI, J. E., SANDINE, W. E., and ELLIKER, P. R., *J. Bacteriol.*, 1965, **89**, 937.
6. COETZEE, J. N., DATTA, N., HEDGES, R. W., and APPELBAUM, P. C., *J. Gen. Microbiol.*, 1973, **76**, 355.
7. COLLINS, L. B., and THOMAS, T. D., *J. Bacteriol.*, 1974, **120**, 52.
8. CORDS, B. R., and McKAY, L. L., *J. Bacteriol.*, 1974, **119**, 830.
9. CORDS, B. R., McKAY, L. L., and GUERRY, P. G., *J. Bacteriol.*, 1974, **117**, 1149.
10. EFSTATHIOU, J. D., and McKAY, L. L., *Appl. Environ. Microbiol.*, 1976, **32**, 38.
11. EFSTATHIOU, J. D., and McKAY, L. L., *J. Bacteriol.*, 1977, **130**, 257.
12. EGAN, J. B., and MORSE, M. L., *Biochem. Biophys. Acta*, 1965, **97**, 310.
13. ELLIKER, P. R., ANDERSON, A. W., and HANNESSON, G., *J. Dairy Sci.*, 1956, **39**, 1611.
14. EXTERKATE, F. A., *Neth. Milk Dairy J.*, 1976, **30**, 3.
15. FARROW, J. A. E., and GARVIE, E. I., *J. Dairy Res.*, 1979, **46**, 121.
16. GASSON, M. J., and DAVIES, F. L., *Soc. Gen. Microbiol. Quarterly*, 1979, **2**, 87.
17. GILLILAND, S. E., SPECK, M. L., and WOODARD, J. R., *Appl. Microbiol.*, 1972, **23**, 21.
18. HAMILTON, I. R., and LO, C. Y., *J. Bacteriol.*, 1978, **136**, 900.
19. HENGSTENBERG, W., EGAN, J. B., and MORSE, M. L., *Proc. Nat. Acad. Sci.*, 1967, **58**, 274.
20. HENGSTENBERG, W., and MORSE, M. L., *Carbohyd. Res.*, 1968, **7**, 180.
21. HENGSTENBERG, W., PENBERTHY, W. K., HILL, K. H., and MORSE, M. L., *J. Bacteriol.*, 1968, **96**, 2187.

22. HENGSTENBERG, W., SCHRECKER, O., STEIN, R., and WEIL, R., *Staphylococci and Staphylococcal Diseases: Proc. 3rd Inter. Symposium on Staphylococci and Staphylococcal Infections*, ed. Jeljaszewicz, J., 1976, Gustavo Fischer Verlag Stuttgart, NY.
23. HIROTA, Y., *Proc. Nat. Acad. Sci. USA*, 1960, **46**, 57.
24. HIRSCH, A., *J. Dairy Res.*, 1952, **19**, 290.
25. HIRSCH, A., *J. Gen. Microbiol.*, 1951, **5**, 208.
26. HORIUCHI, T., TOMIZAWA, J. I., and NOVICK, A., *Biochem. Biophys. Acta*, 1962, **55**, 152.
27. HUNTER, G. J. E., *J. Dairy Res.*, 1939, **10**, 464.
28. JACOB, F., BRENNER, S., and CUZIN, F., *Cold Spring Harbor Symp. Quant. Biol.*, 1963, **28**, 329.
29. JOHNSON, E. M., WOHLHIETER, J. A., PLACEK, B. P., SLEET, R. B., and BARON, L. S., *J. Bacteriol.*, 1976, **125**, 385.
30. JOHNSON, K. G., and MCDONALD, L. J., *J. Bacteriol.*, 1974, **117**, 667.
31. KASHKETT, E. R., and WILSON, T. H., *J. Bacteriol.*, 1972, **109**, 784.
32. KAYSER, F. H., WUST, J., and CORRODI, P., *Antimicrob. Agents Chemother.*, 1972, **2**, 217.
33. KEMPLER, G. M., BALDWIN, K. A., MCKAY, L. L., MORRIS, H. A., HALAMBECK, S., and THORSEN, G., *J. Dairy Sci.*, 1979, **62**, Suppl. 1, 42.
34. KEMPLER, G. M., and MCKAY, L. L., *Appl. Environ. Microbiol.*, 1979, **37**, 316.
35. KEMPLER, G. M., and MCKAY, L. L., *Appl. Environ. Microbiol.*, 1979, **37**, 1041.
36. KEPES, A., and COHEN, G. N., *The Bacteria*, 1962, Academic Press, New York, 179.
37. KLAENHAMMER, T. R., and MCKAY, L. L., *Dairy Sci.* 1976, **59**, 396.
38. KLAENHAMMER, T. R., MCKAY, L. L., and BALDWIN, K. A., *Appl. Environ. Microbiol.*, 1978, **32**, 45.
39. KUHL, S. A., LARSEN, L. D., and MCKAY, L. L., *Appl. Environ. Microbiol.*, 1979, **37**, 1193.
40. LANGRIDGE, J., *Molec. Gen. Genet.*, 1969, **105**, 74.
41. LARSEN, L. D., and MCKAY, L. L., *Appl. Environ. Microbiol.*, 1978, **36**, 944.
42. LAWRENCE, R. C., THOMAS, T. D., and TERZAGHI, B. E., *J. Dairy Res.*, 1976, **43**, 141.
43. LEBLANC, D. J., CROW, V. L., LEE, L. N., and GARON, C. F., *J. Bacteriol.*, 1979, **137**, 878.
44. LEE, R., MOLSKNESS, T., SANDINE, W. E., and ELLIKER, P. R., *Appl. Microbiol.*, 1973, **26**, 951.
45. MACRINA, F. L., and BALBINDER, E., *J. Bacteriol.*, 1972, **112**, 503.
46. MCDONALD, I. J., *Can. J. Microbiol.*, 1975, **21**, 245.
47. MCKAY, L. L., *Food Technol.*, 1978, **32**, 181.
48. MCKAY, L. L., and BALDWIN, K. A., *Appl. Environ. Microbiol.*, 1978, **36**, 360.
49. MCKAY, L. L., and BALDWIN, K. A., *Appl. Microbiol.*, 1973, **25**, 682.
50. MCKAY, L. L., and BALDWIN, K. A., *Appl. Microbiol.*, 1974, **28**, 342.
51. MCKAY, L. L., and BALDWIN, K. A., *Appl. Microbiol.*, 1975, **29**, 546.
52. MCKAY, L. L., BALDWIN, K. A., and EFSTATHIOU, J. D., *Appl. Environ. Microbiol.*, 1976, **32**, 45.
53. MCKAY, L. L., BALDWIN, K. A., and WALSH, P. M., *Appl. Environ. Microbiol.*, 1980, **40**, 84.

54. McKay, L. L., Baldwin, K. A., and Walsh, P. M., *J. Dairy Sci.*, 1979, **62**, Suppl. 1, 43.
55. McKay, L. L., Baldwin, K. A., and Zottola, E. A., *Appl. Microbiol.*, 1972, **23**, 1090.
56. McKay, L. L., Cords, B. R., and Baldwin, K. A., *J. Bacteriol.*, 1973, **115**, 810.
57. McKay, L. L., Miller III, A., Sandine, W. E., and Elliker, P. R., *J. Bacteriol.*, 1970, **102**, 804.
58. McKay, L. L., Walter, L. A., Sandine, W. E., and Elliker, P. R., *J. Bacteriol.*, 1969, **99**, 603.
59. Molskness, T. A., Lee, D. R., Sandine, W. E., and Elliker, P. R., *Appl. Microbiol.*, 1973, **25**, 373.
60. Molskness, T. A., Sandine, W. E., and Brown, L. R., *Appl. Microbiol.*, 1974, **28**, 753.
61. Morse, M. L., Hill, K. L., Egan, J. B., and Hengstenberg, W., *J. Bacteriol.*, 1968, **95**, 2270.
62. Morse, M. L., Lederberg, E. M., and Lederberg, J., *Genetics*, 1956, **41**, 142.
63. Morse, M. L., Lederberg, E. M., and Lederberg, J., *Genetics*, 1956, **41**, 758.
64. Mundt, J. O., *J. Milk and Food Technol.*, 1970, **33**, 550.
65. Novick, A., and Horiuchi, T., *Cold Spring Harbor Symp. Quant. Biol.*, 1961, **26**, 239.
66. Novick, R. P., *Bacteriol. Rev.*, 1969, **33**, 210.
67. Okamoto, T., and Morichi, T., *Agric. Biol. Chem.*, 1979, **43**, 2389.
68. Okulitch, O., and Eagles, B. A., *Can. J. Res. Ser. B*, 1936, **14**, 320.
69. O'Leary, V. S., and Woychik, J. H., *Appl. Environ. Microbiol.*, 1976, **32**, 89.
70. Postma, P. W., and Roseman, S., *Biochem. Biophys. Acta*, 1976, **457**, 213.
71. Reddy, M. S., Williams, F. D., and Reinbold, G. W., *J. Dairy Sci.*, 1973, **56**, 634.
72. Reeve, E. C., and Braithwaite, J. A., *Genet. Res.*, 1973, **22**, 329.
73. Riley, M., and Anilionis, A., *Ann. Rev. Microbiol.*, 1978, **32**, 319.
74. Sandine, W. E., Radich, P. C., and Elliker, P. R., *J. Milk and Food Technol.*, 1972, **34**, 43.
75. Schifsky, R. F., and McKay, L. L., *J. Dairy Sci.*, 1975, **58**, 482.
76. Sharpe, M. E., *J. Soc. Dairy Technol.*, 1979, **32**, 9.
77. Shipley, P. L., and Olsen, R. H., *J. Bacteriol.*, 1975, **123**, 20.
78. Simoni, R. D., Smith, M. F., and Roseman, S., *Biochem. Biophys. Res. Commun.*, 1968, **31**, 804.
79. Somkuti, G. A., and Steinberg, D. H., *J. Dairy Sci. (Suppl.)*, 1978, **61**, 118.
80. Somkuti, G. A., and Steinberg, D. H., *J. Food Prot.*, 1979, **42**, 885.
81. Thomas, T. D., *Appl. Environ. Microbiol.*, 1976, **32**, 474.
82. Thomas, T. D., *J. Bacteriol.*, 1976, **125**, 1240.
83. Thompson, J., *J. Bacteriol.*, 1978, **136**, 465.
84. Thompson, J., *J. Bacteriol.*, 1979, **140**, 774.
85. Thompson, J., and Thomas, T. D., *J. Bacteriol.*, 1977, **130**, 583.
86. Warren, R. A., *Can. J. Microbiol.*, 1972, **18**, 1439.
87. Wohlhiete, J. A., Falkow, S., Citarella, R. V., and Baron, L. S., *J. Mol. Biol.*, 1964, **9**, 576.
88. Yawger, E. S., and Sherman, J. M., *J. Dairy Sci.*, 1937, **20**, 83.

Chapter 6

NEW DEVELOPMENTS IN THE RAPID ESTIMATION OF MICROBIAL POPULATIONS IN FOODS

J. M. WOOD and P. A. GIBBS

*Leatherhead Food Research Association,
Leatherhead, UK*

SUMMARY

Routine methods for the estimation of microbial populations in foods have changed little over recent decades. There has, however, in the last few years been much research into more rapid and less labour intensive methods of estimation. New methods of estimation which the authors feel will come into routine use in food microbiology have been considered.

These methods include direct cell counts by epifluorescent microscopy, metabolically based techniques, e.g. impedimetric and radiometric estimations and techniques which directly estimate constituents of microbial cells e.g. ATP or endotoxins.

The separation of microorganisms from food has been considered as an adjunct to more rapid estimation and some implications of the use of metabolically based estimation techniques in food microbiology have been examined.

INTRODUCTION

In the time scale of the development of microbiological methods almost all of the 'rapid methods' might be considered as new developments because few, as yet, are used in routine analyses. This is not to say that there has been a lack of interest in the field. The proceedings of international symposia[1,2]

183

demonstrate a wealth of interest at the research level over a broad range of methods. Much of this interest has arisen from the medical area, however food microbiologists, many of whom work with perishable products, are also motivated to develop more rapid methods.

The objective of this chapter is to describe some of the developments which are emerging in the rapid estimation of microbial populations in food and which we feel are likely to find routine use in food microbiology. A further objective is to consider the impact which some of these methods may have on our future approach to the estimation of microorganisms in foods.

CELL AND COLONY COUNTING METHODS

The direct microscopic count is the original rapid method and when applicable to a food is also simple to perform. The use of this method in food microbiology is limited by several factors; low sensitivity, food materials which obscure the field and of course the fatigue associated with long periods at the microscope. Recent developments by Pettipher and his colleagues[3,4] appear to have overcome these problems in the estimation of microorganisms in raw milk and some dairy products.

These workers have developed and evaluated a rapid direct epifluorescent filter technique (DEFT). In this, clumps of microorganisms collected or recovered from the product by filtration are stained *in situ* with a fluorescent dye and the clumps counted. This estimate which is known as a membrane clump count (MCC), can be made in 20 mins.

Membrane filtration and epifluorescent microscopy are not new techniques, the essence of the DEFT is the method developed to permit the filtration of useful quantities of product. The major particles in raw milk which block filter pores are fat globules, somatic cells and aggregates of protein. In the DEFT these are dispersed by a combined treatment using proteolytic enzymes and detergents. The treatment for raw milk is as follows: 0·5 ml of a 20 % solution of a crude trypsin preparation and 2·0 ml of 0·5 % Triton X-100 detergent are added to 2 ml of milk and the mixture held at 50 °C for 10 mins. After this period the warm mixture is filtered through a polycarbonate 0·6 μm filter which is then stained with an Acridine Orange/Tinopal stain. Modifications of this procedure, e.g. differences in the concentration of detergent and/or the period of incubation, permitted the filtration of a range of dairy products, e.g. butter, cream, pasteurised milk and UHT milk.

The results prescribed by Pettipher and Rodrigues[3] show a close

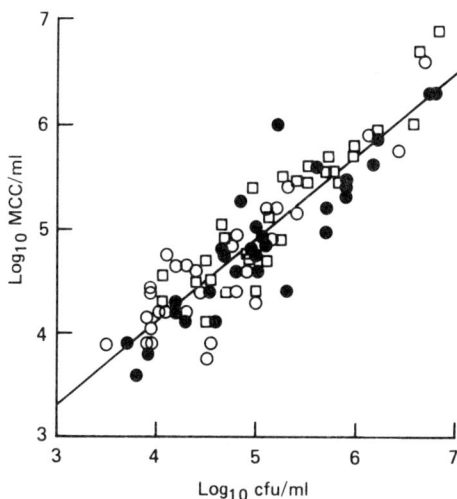

FIG. 1.   Correlation between membrane and plate counts for 30 samples each of exfarm ($\bigcirc$), bulked tanker ($\bullet$), and silo milks ($\square$). Line represents fitted regression line ($y = 0{\cdot}89 + 0{\cdot}8x$).

agreement between their technique and the colony count over the range $10^4–10^7$ cfu ml^{-1} (Fig. 1).

This method in the form first reported still had the potential disadvantage of fatigue caused by long periods of microscopic observations. More recently, further developments have been made in which the counting procedure has been automated by the use of an image analyser and it appears that the complete technique could have an important role in dairy microbiology. A collaborative trial of the technique has now been carried out in several UK dairy laboratories; the results of this trial indicate good 'inter-laboratory' reproducibility of results.

Colony counting methods i.e. plate counts, are by definition not rapid because of the incubation periods necessary for the growth of colony forming units to colonies. However, colony count is still the most commonly used method of estimating microbial populations and most microbiologists are alert for possibilities of reducing the time and effort involved in these analyses.

Over the last ten years there have been many developments in the mechanisation and miniaturisation of colony count techniques. Methods such as the Agar Droplette Technique,[5] loop dilution technique and machines such as the Spiral Plate Maker and the many types of automatic colony counters are now well established in food microbiology. It is interesting though to note the time required for one of these methods to 'take off' in food microbiology since this may give an idea of the period which might elapse before some of the currently published developments are in common use. The Spiral Plate Method was developed in America and published in 1973 by Gilchrist *et al.*[6] Evaluations of the use of the method for estimation of microorganisms in foods appeared some years later.[7-9] A commercial instrument became available in the UK in 1978, and whilst the method is commonly used in the UK (and Europe) only in 1981 has it become an accepted AOAC method.[10]

NON-COUNTING METHODS

In analytical terms the growth of microorganisms to colonies on gelified media is a powerful amplification procedure which permits 'naked-eye' enumeration of microorganisms. In this, the most practiced enumeration technique in microbiology, convenience is gained but time is lost.

The desire to obtain results more rapidly, and preferably with less labour has usually led to non-colony counting techniques. The light scattering techniques which have long been used to estimate microbial populations in clear liquid media (i.e. in model systems) are not applicable to foods. However, the use of dye reduction tests in the dairy industry has demonstrated the potential of non-counting methods in food microbiology.

The dye reduction tests provide an estimate of microbial content by the observation of changes brought about in the medium by the metabolic activities of populations of viable microorganisms. Such non-counting methods which take metabolic activity as a basis for estimation, have provided some of the more interesting developments in rapid estimation.

In these developments very sensitive analytical instruments are used to detect changes in, for example, electrical characteristics of a medium, evolution of gas or heat production caused by growing microorganisms. The time to detect such changes is inversely related to the number of organisms originally present in the inoculum (see Fig. 3 for impedance) and the time to detection is usually much shorter than that required for the growth of colonies.

Several commercially available detection systems have been shown to be capable of producing rapid estimates of microorganisms in foods. The most promising results have been obtained with impedimetric and radiometric techniques.

Impedimetric Estimation of Microbial Populations

Electrical impedance can be defined as the resistance to the passage of an alternating current. When microorganisms grow in a nutrient medium their metabolic activities and/or products change the impedance of the medium. Such changes can be continuously monitored by passing a small electric current through the medium and comparing the values obtained against an uninoculated control medium or against a predetermined impedance value.

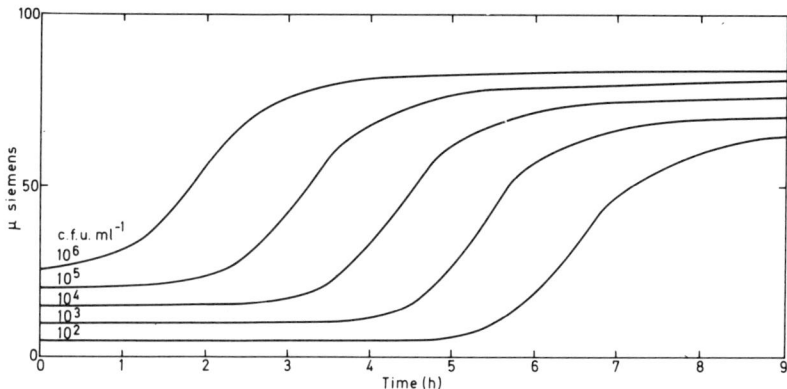

FIG. 2.   Impedance changes with time for *Escherichia coli* grown in PPLO broth at 37°C.

A comprehensive description of the physics of impedimetric estimation may be found in the paper by Richards *et al.*[11] As growth proceeds the impedance of a microbial culture (usually) decreases and this change is first detected when the concentration of organisms exceeds a 'threshold' of about 10^6 cfu ml^{-1}. The time to reach this threshold is a function of the initial concentration of organisms and their specific growth kinetics; by measuring the time to detection the concentration of organisms originally present can be determined (Figs 2 and 3).

This technique is far from new. The impedimetric responses of microorganisms were first reported by Stewart in 1899.[12] The technique reappeared in the early 1970s when multi-channel impedance instruments with continuous recording of signals became available. These instruments were first tried in clinical microbiology but in about 1975 began to be considered for the estimation of microorganisms in foods. Results obtained with this technique began to appear in 1977–1978. Hardy *et al.*[13] looked at frozen vegetables, Cady *et al.*[14] at milk and Wood *et al.*[15] at raw meat and frozen vegetables. There followed an interruption in the supply of

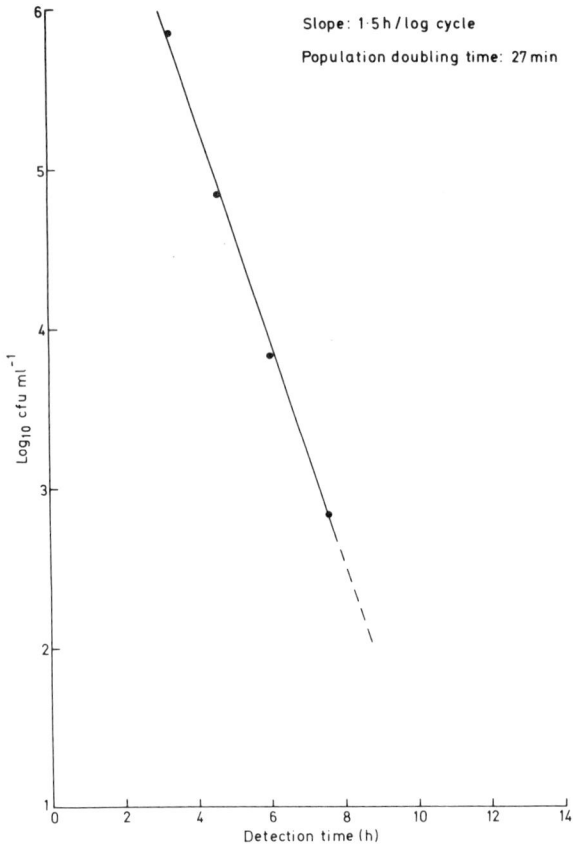

FIG. 3.   Relationship between initial viable count and detection time for *E. coli* in brain–heart infusion broth at 37 °C.

instruments and this caused a pause in this work but with the resumption in the supply of instruments many food microbiologists have begun to evaluate this technique.

Two manufacturers currently produce impedance monitoring instruments for microbiological analyses; Malthus Instruments Ltd of Perth, Scotland, and Bactomatic Inc. of Princeton, New Jersey, USA, and Marlow, Bucks, England.

Bactomatic Inc. currently produce two instruments. The smaller instrument can monitor up to 32 samples simultaneously and will continuously record the responses on a chart recorder. The larger instrument can monitor up to 120 samples simultaneously and has been specifically designed for quality control work. When samples are placed in the machine the sample description together with the acceptable detection time are fed into the instrument memory via a microprocessor keyboard. As incubation proceeds a computer regularly scans the samples for changes in impedance and the time to reach the detection threshold for impedance change is shown on a video display unit. If this time is less than the acceptable detection time (i.e. more organisms are present than are acceptable for that sample) this fact is emphasised by presenting the figure against a background of different colour from that of the acceptable detection times.

These instruments can accept widely different sample volumes but most commonly 2 ml samples have been used in the presterilised and disposable 16 cell modules supplied by the manufacturer. These fit into a small air incubator in the instrument.

The Malthus instruments were developed from the work of Dr G. Hobbs and his colleagues at the Torry Research Station in Aberdeen, Scotland. There is an eight channel machine which continuously records responses onto a chart recorder and a more sophisticated 120 channel instrument which is computer controlled and is instructed via a keyboard. This instrument 'memorises' all of the monitored impedance changes and can be interrogated at any time. A complete record of the impedance change with time can be obtained for any sample either as a numerical print-out, or graphically on the video display unit, or on paper.

These instruments accept samples in glass containers into which an autoclavable electrode assembly is placed. Incubation of the containers is carried out in a water bath.

These are brief and incomplete descriptions of these instruments but serve to indicate the introduction of computer and microprocessor controlled instrumentation into food microbiology. The fact that such instruments were developed as rapid methods should not obscure the other advantages of this instrumentation. The capacity to analyse simultaneously large numbers of samples, with a minimum of sample preparation and with continuous monitoring and recording of data are significant developments in their own right. Microbiologists will doubtless avail themselves of these labour-saving advantages in many fields of investigation.

In food microbiology, impedimetric techniques have been investigated most commonly for the rapid estimation of 'total viable microorganisms'. That is, impedimetric detection times have been compared with 'total colony counts' for particular foods. Hardy et al.[13] compared the two techniques for the estimation of microorganisms in frozen vegetables. They

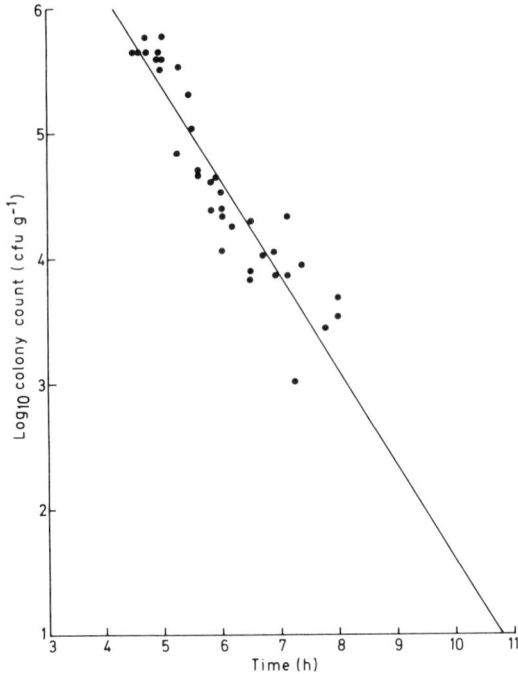

FIG. 4. Relationship between detection time and bacterial colony count at 35 °C for 36 samples of frozen vegetables (retail and laboratory-abused samples) determined using Bactometer 32.

found that the technique was rapid, for example 10^5 cfu g^{-1} were detected in 4·5 h and that the agreement between the two methods in classifying samples about specified levels of colony count was high. In our own experience 10^5 cfu g^{-1} in frozen vegetables were detected in 5·2 h and agreement between the detection time and colony count was similar to that found by Hardy et al.[13]

The relationship between impedimetric detection time and colony count for retail samples of frozen vegetables is shown in Fig. 4. This demonstrates the relationship over a one thousandfold range of colony count. When this method was used to monitor actual production of frozen peas the data shown in Fig. 5 were obtained. These results which were obtained on five consecutive days of production show detection times of about 5 and 7 h for cfu g^{-1} levels of 10^5 and 10^4 respectively. Shorter detection times than 5 h would indicate that the colony count had exceeded 10^5 cfu g^{-1}. The production of frozen peas is geared to the harvesting of the peas and the

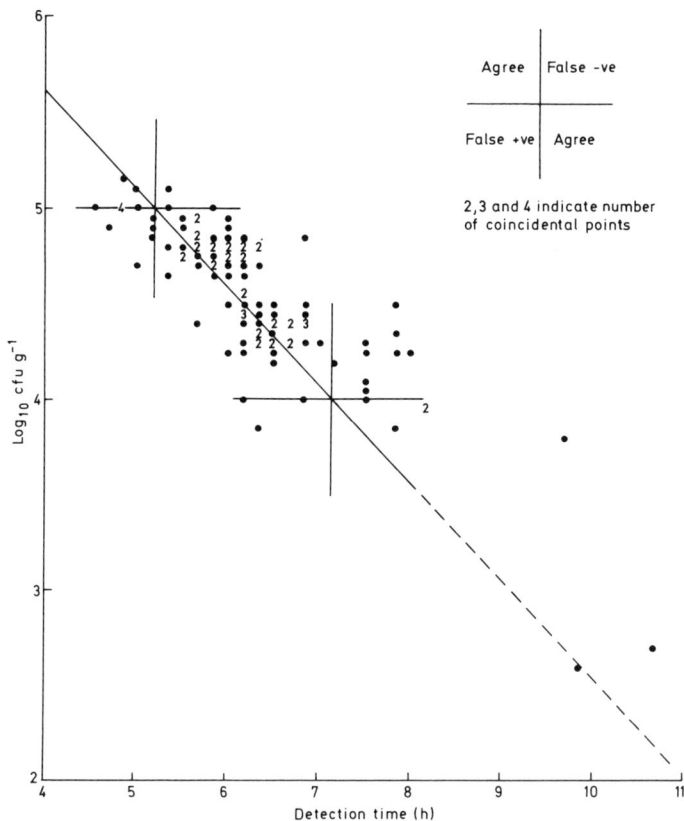

FIG. 5.   Distribution of data about the derived regression line for 117 samples of frozen peas tested by Bactometer compared with Spiral Plate counts at 30 °C for 48 h.

production line runs continuously for long periods. The advantage of rapid monitoring here is that it is possible within hours rather than days, to estimate the levels of microorganisms and from this information make decisions about the cleaning of the production line.

The relationship between impedimetric detection time and colony count for raw meat is shown in Fig. 6 and the percentage agreement between the methods in classifying the samples about specified levels of colony count are shown in Table 1. The calculation of results in agreement was made by the method reported by Hardy et al.[13] The specified levels of colony count are shown in Fig. 6. At the point where these values intersect the regression line the value of impedimetric detection time for this colony count level is

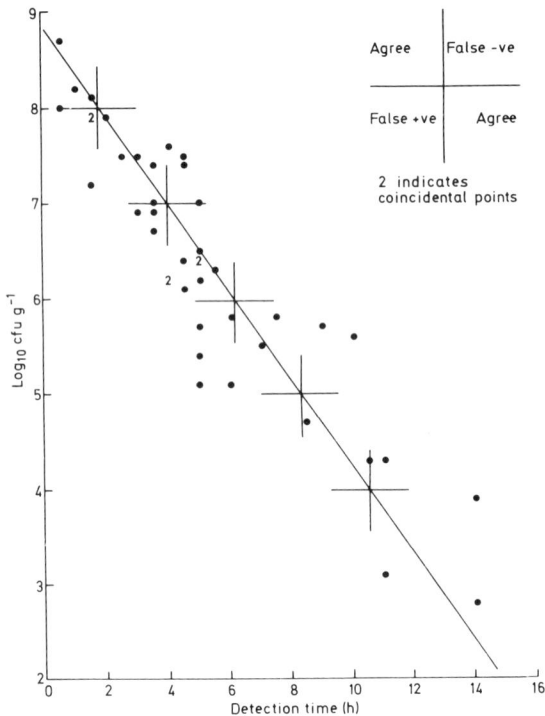

FIG. 6.   Distribution of data points about the derived regression line for 43 raw meat samples tested by Bactometer compared with Spiral Plate counts at 30 °C for 48 h.

interpolated from the graph. This value is the 'cut off time' for that colony count level. A grid is then constructed at this point and the data points assigned as follows:

Results in agreement
   (a) Samples having detection times shorter than cut off time with counts greater than the specified level.
   (b) Samples having detection times longer than the cut off time with counts lower than the specified level.

Results not in agreement
   (a) Samples having detection times shorter than the cut off time but with counts lower than the specified level.
   (b) Samples having detection times longer than the cut off times but with a colony count higher than the specified level.

The level of agreement found here indicates that for the microbiological grading of raw meat the impedimetric method would produce substantially the same results as the colony count technique. The data shown in Fig. 6 were produced in 1978 and it is interesting to note that Bell[16] found similar 'cut off' times for raw meat in his study, i.e. 6.0 h for 10^6 cfu g^{-1} and 7.5 h for 10^5 cfu g^{-1}.

TABLE 1

AGREEMENT BETWEEN COLONY COUNT AND IMPEDIMETRIC ESTIMATION TECHNIQUES FOR THE CLASSIFICATION OF 43 RAW MEAT SAMPLES ABOUT SPECIFIED LEVELS OF COLONY COUNT

Specified level (cfu g^{-1})	10^8	10^7	10^6	10^5	10^4
Cut off time (h)	1·8	4·0	6·2	8·2	10·6
% Agreement	93	88	93	95	98

This use of impedence techniques for the determination of total colony count requires the construction of a calibration curve for each type of product in which qualitatively different microbial floras may occur. In addition, since the impedimetric signal is related to the growth kinetics of the organisms growth conditions should be optimised. An example of the effect of product inhibition of growth is shown in Fig. 7 for ice cream mix. Normally it is assumed that when more product is presented to the machine, the more organisms will be present and therefore the shorter will be the detection time. It can be seen from Fig. 7 that this is not always the case. We have presumed that the higher concentration of this product contains sufficient low molecular weight materials to restrain growth and that this restraint is relieved when the product is further diluted.

Currently the impedimetric estimation of total microbial populations in foods is under active investigation by food microbiologists. There are investigations of other applications of impedance technique, for instance into the estimation or detection of specific groups of organisms, viz. coliforms, salmonellae, yeasts and moulds.

The work with moulds is proving interesting from several viewpoints. Firstly our results at Leatherhead (Williams and Wood, unpublished observations) have shown that in media commonly used for the estimation of moulds by colony count, common food moulds produce an *increase* in impedance rather than the decrease observed for bacteria. This is true whether the organisms are grown in the gelified media or in media of the same composition except for agar. This does not prevent the impedimetric estimation of these organisms. It can be seen from the impedimetric

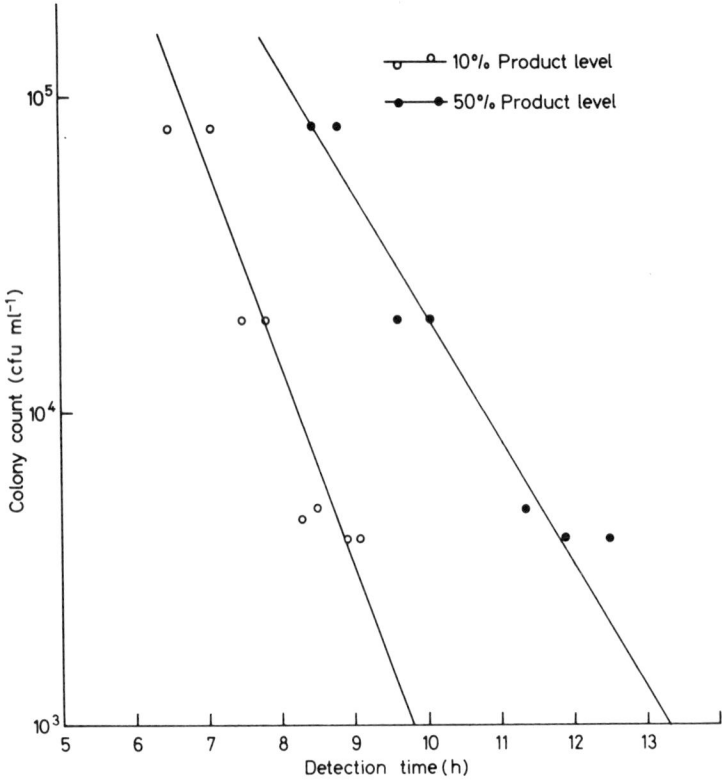

FIG. 7.   The impedimetric estimation of bacteria in ice cream mix diluted into tryptone soya broth with 0·1 % yeast extract.

responses produced (Fig. 8) that an inverse linear relationship exists between inoculum level (colony count) and the impedimetric detection time (Fig. 9).

A problem which arises in this type of comparison is that there is no very precise method for the estimation of moulds in foods. The Howard Mould Count (direct microscopic count of mycelial fragments) may vary with the treatment of the product and in colony counts the colony forming unit or mould propagule may represent a single spore—or a mass of mycelium. The data shown in Figs 8 and 9 were obtained using a spore preparation to provide a uniformly dispersed inoculum; however, this is unlikely to be the case in foods. It is at least possible therefore that for the estimation of moulds in foods metabolically based estimates such as impedance may

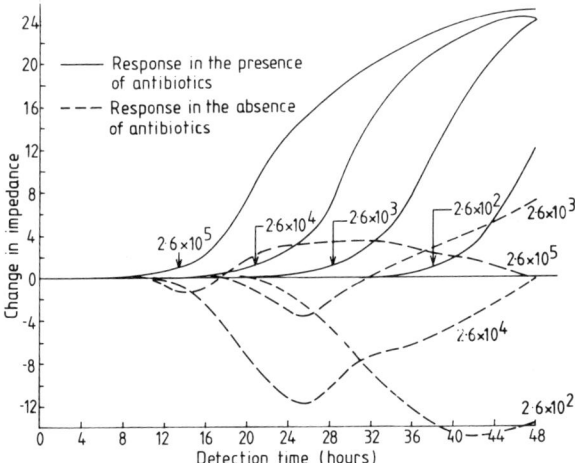

FIG. 8. Detection times for dilutions of mouldy cheese with antibiotics (chloramphenicol 100 ppm, oxytetracycline 100 ppm). Data from Williams and Wood, unpublished.

provide more meaningful estimates of mould biomass present than do colony counts.

An interesting observation was made during the formulation of a selective medium for the impedimetric estimation of moulds in foods. Antibiotics added to the medium to inhibit bacteria did so but also caused a delay in the impedimetric detection of the mould. This demonstrated that

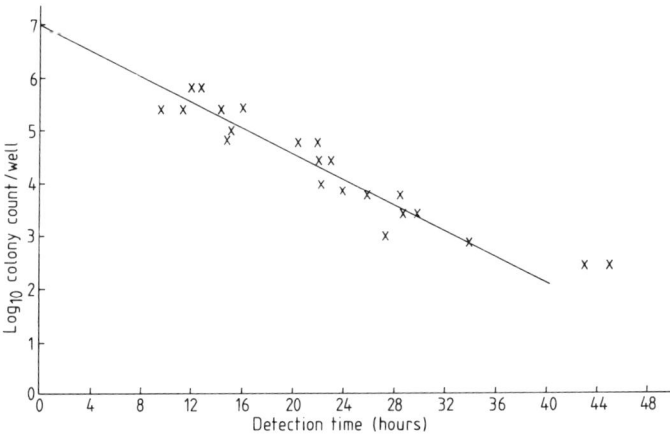

FIG. 9. Relationship between detection time and colony count for *Aspergillus ochraceus* (M209). Data from Williams and Wood, unpublished.

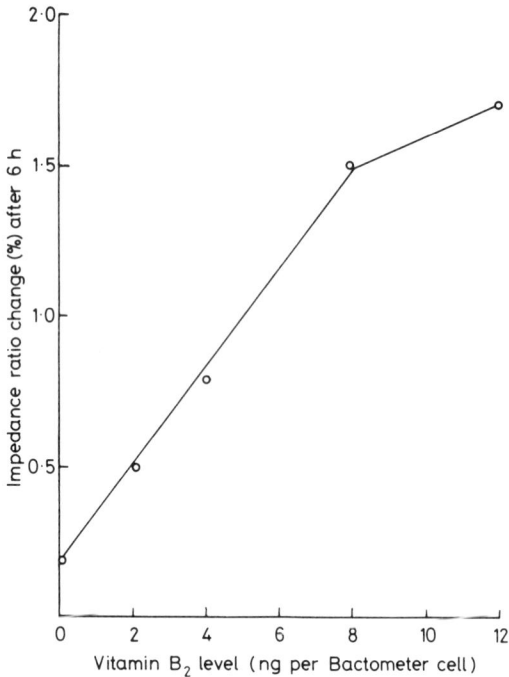

FIG. 10.   Impedimetric assay of riboflavin using *Streptococcus faecalis* (NCIB 7432).

the antibiotics were imperfect selective agents and also provided a rapid and quantitative measure of the extent of this effect. Thus the formulation and testing of such selective media can be carried out more quickly than that for colony counts.

Other applications of impedance measurements within food microbiology are the rapid assessment of the activity of starter cultures before use in fermented products, and in those analyses in which the growth of organisms is used to assay the concentration of a substance in a food e.g. vitamins and antibiotic residue assays. We have demonstrated the feasibility of the impedimetric assay of vitamins (Fig. 10) and there appears to be no reason why antibiotics e.g. in milk, should not be estimated by the inverse procedure, i.e. inhibition of growth/impedance signal.

Radiometric Estimation

Radiometry involves the measurement of $^{14}CO_2$ produced by microbial metabolism of ^{14}C- labelled substrates incorporated in a culture medium.

The time taken to detection has an inverse linear relationship to the initial number of organisms in the sample. The instrument most commonly used is the 'Bactec' (Johnston Laboratories, Maryland, USA) in which the head-space gases above the labelled medium, previously inoculated with the sample, are tested at intervals. Positive results for higher concentrations of microorganisms in foods can be obtained within the working day (Rowley et al.).[1,17]

Unlike the impedance instruments the 'Bactec' does not provide a continuous monitoring system. It is essentially a reader into which the culture bottles are placed by the operator. Incubation is carried out in a conventional incubator prior to testing and culture bottles placed in the instrument are automatically sampled at a rate of 60 per hour in the 'Bactec 460'. There is also a simpler, manual instrument—the Bactec 301B. The overall working capacity of these instruments will obviously depend upon the pattern of work presented to them.

The radiometric estimation and detection of microorganisms has been most commonly used for clinical samples, however, the radiometric estimation of microorganisms in foods has also been reported. Previte[18] considered the estimation of pathogenic organisms in food and also demonstrated the radiometric estimation of *Salmonella typhimurium* and *Staphylococcus aureus*. 10^4 cfu ml^{-1} of *S. typhimurium* were detected in 3 or 4 h depending upon the specific activity of the labelled substrate.

Hatcher[19] used the 'Bactec 301' system to estimate the number of organisms in frozen orange juice and reported that concentrations of 10^4 ml^{-1} yeast, *Leuconostoc* and *Lactobacillus* were detected in 6, 7 and 10 h respectively. Of 600 orange juice samples examined 44 positives (10^4 cfu ml^{-1}) were detected in 12 h, 41 of which were positive in 8 h. Rowley et al.[17] investigated the use of radiometry as a rapid screening method for cooked, frozen foods. Seventy-five per cent of a wide variety of the food samples were correctly classified as acceptable ($\leq 1 \times 10^5$ cfu g^{-1}) or unacceptable ($> 1 \times 10^5$ cfu g^{-1}) within 6 h, and a maximum of 5 of the 404 samples tested were incorrectly classified. These reports indicate that radiometry is a promising technique for use in food microbiology. A further development has recently been reported by Stewart and Eyles[20] in which salmonellae are detected by their ability to produce gas from radiolabelled dulcitol. It was observed that this reaction was suppressed by salmonella polyvalent 'H' antiserum and the ratio of gas produced in the absence of antiserum to that produced in the presence of antiserum was used as the criterion of detection. This is an intriguing report not only because it may offer a novel method of detecting commonly sought food poisoning organisms, but also because it appears to offer a novel method of observing the interaction of a microorganism with an antibody.

As mentioned above, the greatest use of radiometric analyses in microbiology has been in the clinical field where the use of radioisotopes has long been accepted. Food manufacturers have expressed some hesitancy in using radioactive materials in laboratories close to food production sites just as they are reluctant to handle pure cultures of known pathogens in their laboratories. However, in laboratories well separated from production the problem of cross contamination does not arise and there would appear to be little reason for food microbiologists to deny themselves the advantages inherent in the analytical use of radioisotopes.

Very Rapid Techniques
With the exception of direct microscopic observation the most rapid methods of estimating microbial populations are those which measure a structural or metabolic component of the microbial cell. Several methods have been demonstrated.

The 'Limulus' assay is based upon the observation by Levin and Bang[21] that minute amounts of endotoxin from the outer membrane of Gram-negative bacteria will coagulate a lysate of amoebocytes from the horseshoe crab (*Limulus polyphemus*). Assays of microorganisms based upon this observation have been reported by Jorgensen *et al.*[22] and Coates.[23] Coates found that for all of the species examined, $5 \times 10^2 \, \text{cfu ml}^{-1}$ could be detected in 1 h and that this number of organisms represented 0·05 ng of endotoxin. Jay[24] found the Limulus Lysate Test to be an excellent, simple and rapid test of the microbial quality of refrigerated ground beef. In this test endotoxins were extracted from bacteria in the beef by simply shaking the meat in pyrogen-free distilled water. Serial dilutions of the extract were prepared and an aliquot of a standardised Limulus preparation (Difco Pyatest Reagent) added. After incubation at 37 °C for 1 h the tubes were observed for gelation and the 'titre' recorded as the reciprocal of the highest dilution producing a gel or a clot. This titre correlated well with the level of Gram-negative microorganisms in beef. Based upon this study Jay has recommended that a 0·1 ml inoculum from a 10^3 dilution of good quality ground beef should produce a negative lysate test and thus serve as an additional rapid screening test for meat microbial quality.

Two other methods demonstrated to produce very rapid estimates of microbial populations are the chemiluminescence and the bioluminescence methods. Both of these employ photometers to measure the light produced by the reaction of microbial cell materials with added reagents.

The chemiluminescence assay is based upon the catalytic effect of porphyrins (haematin) on the chemiluminescent reaction of luminol (5-amino-2,3,-dihydro-1,4-aminophthalate) in the presence of hydrogen peroxide. Extremely small amounts of porphyrins (10^{-11} g) can be detected in cell free extracts of bacteria and yeasts and Oleniacz et al.[25] found that in general the lower level of sensitivity of this assay for bacteria was between 10^3 and 10^4 cells. This very sensitive assay was found by Strange[26] to give poorly reproducible results. However, Picciolo et al.[27] demonstrated that non-specific reactions due to metal ions or to soluble porphyrins could be minimised making the reaction more specific for intact bacterial cells.

The bioluminescence assay of microbial populations is based upon the reaction of adenosine triphosphate (ATP) from microbial cells with the luciferin–luciferase enzyme system from fire-fly tails. The light produced from the reaction permits the estimation of ATP down to picogram levels (10^{-12} g). The use of this reaction for the rapid estimation of microbial populations was reported by Levin et al. in 1967[28] and its use for the estimation of microorganisms in foods by Sharpe et al.[29] in 1970. Sharpe and his coworkers found that the method gave good results for pure cultures of organisms and reported a value of 1 femtogram (10^{-15} g) per cfu for vegetative cells. However, attempts to estimate microbial populations in a range of foods were unsuccessful due to high levels of non-microbial ATP in the foods. These workers concluded that whilst bioluminescence provided good estimates of microbial populations in pure cultures of microorganisms the method could not be applied to food materials unless a system for the separation of microorganisms from food could be devised. Similar conclusions were reported more recently by Baumgart et al.[30]

There are in fact two ways of resolving the interference problem of non-microbial ATP in assays of microbial ATP. In these assays the ATP must be extracted from the microbial cells prior to the actual measurement. If, before this point, the non-microbial ATP can be destroyed then the assay becomes specific for microbial ATP. This approach has been used successfully in the clinical field for the rapid estimation of microorganisms in urine.[31]

This approach has also been proposed for the estimation of yeasts in orange juice by Vanstaen[32] who has reported that after treatment of orange juice with a nucleotide releasing agent which attacks only the plant cells (NRS reagent, Lumac, Holland), the non-microbial ATP released can be destroyed by incubation with ATPase (Somase, Lumac) for 45 min at 37 °C. The microbial ATP is then released from the yeast cells with a stronger nucleotide releasing agent (NRB reagent, Lumac) and it is proposed that

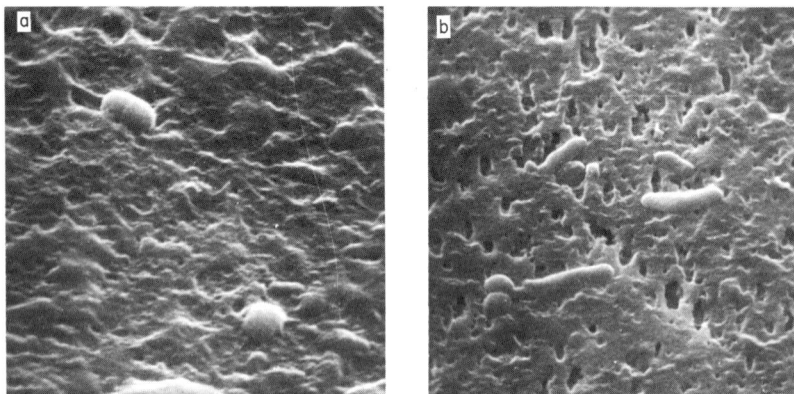

Fig. 11. The separation of microorganisms from meat by the use of cation exchange resin. (a) Microorganisms filtered from supernatant of minced beef homogenate (× 5700). (b) Microorganisms filtered from supernatant solution of minced beef homogenate after passage through cation exchange resin (× 6000).

this microbial ATP can then be assayed before appreciable enzymic destruction occurs.

The alternative approach to dealing with non-microbial ATP in foods is to physically separate the microorganisms from the food or to employ a combination of physical and enzymic methods.

In our own work at Leatherhead (Wood and Stannard, unpublished observations) we have investigated the physical separation of microorganisms from raw meats sufficient to permit the rapid estimation of the meat flora. Microorganisms were separated from meat homogenates by a simple three stage process. Centrifugation to remove coarse food particles; stirring with ion exchange resin to remove protein and then

TABLE 2

THE SEPARATION OF MICROORGANISMS FROM RAW MEAT

1. Homogenise 10 g of meat in 90 ml of phosphate–citrate buffer (0·05M; pH 5·8).
2. Briefly centrifuge an aliquot of the homogenate (holding time of 2000 × g for not more than 10 secs).
3. Remove 10 ml of supernatant from centrifugation to a 50 ml beaker and stir for 2 min with 2·5 g of cation exchange resin (Bio-Rex 70, Bio-Rad Laboratories, Richmond, USA) previously equilibrated with buffer mentioned above. Leave to settle for 3 min.
4. Filter 5 ml of supernatant from resin treatment through 0·22 μm filter.
5. Extract filter with NRB reagent (Lumac) and estimate ATP.

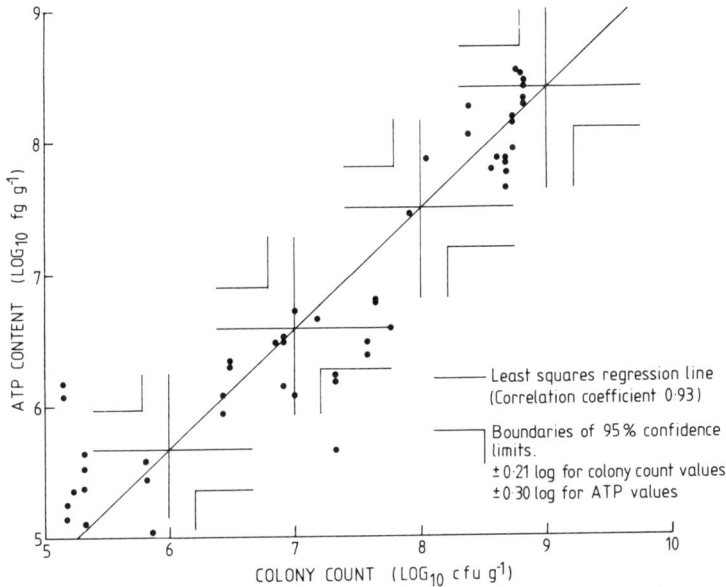

FIG. 12.   Agreement between the two methods at different levels of colony count for beef.

filtration through $0.22\,\mu m$ membranes to remove soluble materials. This last stage also collects and concentrates the organisms. The whole filter membrane is then extracted for ATP. The details of this procedure are outlined in Table 2 and the degree of separation of microorganisms from meat components which can be achieved is illustrated in Fig. 11. This shows the appearance of organisms on filters 'before and after' the resin treatment, before the treatment the filter is clogged by non-microbial material whilst after the treatment only microorganisms are seen. A single assay can be carried out in 20–25 mins.

Some of the results obtained for raw meats are shown in Fig. 12. These show a linear relationship between $\log_{10}$ ATP and $\log_{10}$ colony count with a correlation coefficient (for the beef data) of 0.93. Similar results were obtained with raw lamb and raw pork. The lower limit of sensitivity in this work was governed by the detection limit of the photometer used (Luminometer, LKB Ltd) which was about 1×10^{-12} g ATP. In this system the lower limit of detection corresponded to 10^5 cfu g^{-1} of meat. This was considered sufficiently low for the assay of raw meats but more sensitive instruments are available which will detect about 1×10^{-14} g of ATP. Our own estimates on organisms extracted from raw meat indicate an average value for ATP cfu^{-1} of 5×10^{-16} g (0·5 femtograms). This suggests that in theory at least it is possible to detect about 200 cells. The realisation of this theoretical value depends upon many factors: separation of micro-organisms from the food; separation from non-microbial ATP; concentration of the separated cells and delivery of all of the extracted ATP into the assay vial. It will be interesting to see if this can be achieved; for many practical purposes such levels of organisms are either irrelevant or will be assayed by employing a short period of growth so that the microbial population can increase to more easily detectable levels prior to the assay.

A further example of the separation of non-microbial from microbial ATP is provided by the rapid estimation of yeasts in orange juice (Stannard and Wood, unpublished data). Orange juice contains considerable amounts of non-microbial ATP which completely masks any microbial ATP present. However, if dilute juice is centrifuged at $2000 \times g$ for 10 min, yeasts and cellular plant material are precipitated and the soluble ATP may be decanted in the supernatant solution. The precipitate is then treated with a mild nucleotide releasing agent such as NRS (Lumac) and apyrase (Somase, Lumac) for 45 min at 37 °C to destroy residual non-microbial ATP. The treated precipitate is then resuspended in water and recentrifuged to remove apyrase prior to release of ATP from yeasts with the stronger nucleotide releasing agent (NRB, Lumac). The results of this procedure for orange juice artificially inoculated with yeast (Fig. 13) demonstrate a good basis for a rapid assay and we are in the process of testing this system with a range of spoilage yeast isolates.

The above work indicates that whilst methods of estimation of microbial populations in solid foods by measurement of microbial ATP will be more complex than may initially have been hoped for, this field is developing and beginning to yield interesting prospects for rapid estimation.

We have only briefly mentioned the ATP measuring instruments and reagents with which we are familiar. The development and availability of both instruments and reagents has been of major importance in the field of ATP measurement. Since the early work of Sharpe et al.[29] with foods, new sources of both instruments and reagents have appeared. We have carried out much of our work with the LKB luminometer and LKB reagents which have proved both robust and reliable. We have found in direct comparisons that this instrument is less sensitive than the more expensive machine produced by Lumac although both are simple to use and have proved very reliable in our hands. Both of these instruments are available in Europe and others available from American companies have been catalogued by Goldschmidt and Fung[33] in their report of instruments available for microbiological analysis. One aspect of the reagents proffered for ATP

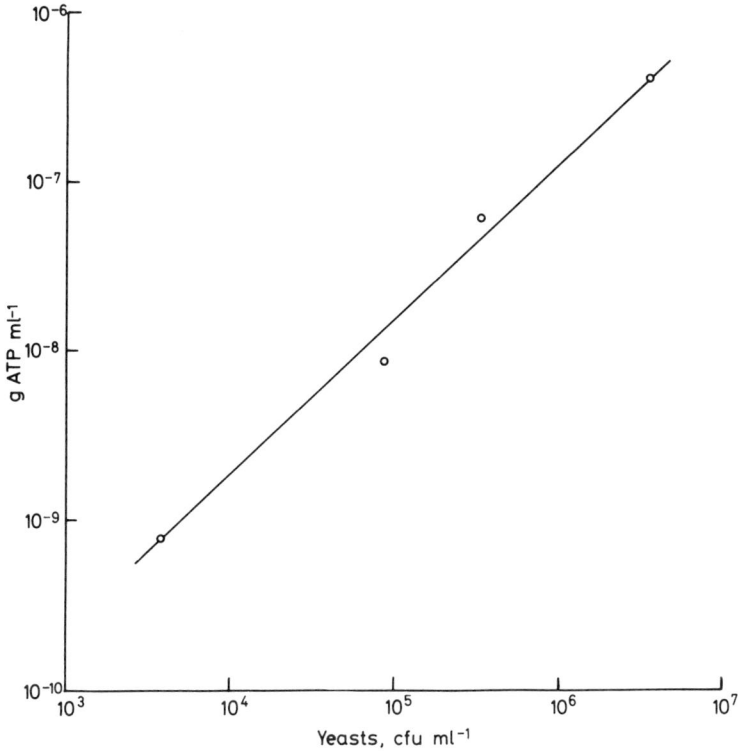

FIG. 13.   Relationship between ATP and cfu ml^{-1} for *Sacch. cerevisiae* in inoculated orange juice.

analysis which may be considered underdeveloped, is that of extractants for release of ATP from cells. Many extractants can be used[34] but some, such as the strong acids are less attractive from the viewpoint of convenience and safety. We have successfully used boiling solutions of Triton X-100 (0·1 % w/v); more recently we have also used the Lumac reagents NRS (for somatic cells) and NRB (for microbial cells). These appear to work well but are of unknown (by us) composition. This is a less than satisfactory situation from the scientific viewpoint but will probably be resolved as work continues.

THE SEPARATION OF MICROORGANISMS FROM FOODS

A well developed facet of most natural sciences is the methodology by which workers separate from naturally occurring systems, that part of the system which they wish to investigate. An impressive example of this is the range of methods used by biochemists to extract specific macromolecules from tissues. In contrast, microbiology has few techniques for the physical separation or collection of microorganisms from solid materials. The major techniques, filtration, centrifugation and flocculation are not commonly applicable to food homogenates and other techniques demonstrated for pure cultures e.g. electrophoresis and counter-current distribution have not been investigated for foods.

There has in fact been little motivation to investigate such techniques since in general foods do not interfere with the traditional cultural techniques. In the search for more rapid methods of estimation several valuable aspects of separation procedures become apparent. Firstly, such methods can remove non-microbial materials which interfere with rapid assay systems, e.g. non-microbial ATP or porphyrins in luminescence assays. Secondly they can permit the concentration of microorganisms from food by removing materials which prevent filtration or which sediment with the organisms in centrifugation. This would in turn permit greater sensitivity to be achieved in the rapid assays since more cells can be presented to the estimation system. It would also shorten the detection period required in non-counting systems such as impedance measurement and radiometry . A more distinct but intriguing possibility is that the physical separation of specific groups of organisms might supplement or even supplant cultural enrichment techniques.

The value of separation in rapid methods is illustrated by the DEFT technique,[3,4] the critical requirement in the technique was to separate the microorganisms from other materials in milk so that they could be concentrated and observed. From this viewpoint the treatment with detergent and protease was in fact a development in filtration–separation. In this context the work of Peterkin and Sharpe[35] is of interest. They have also found that the amounts of various dairy products which can be filtered are greatly increased by pretreating the products with detergent and a protease. The difference in the findings is that whilst Pettipher et al.[3] have used Triton-X-100 which kills the cells, Peterkin and Sharpe have used Tween 80 which does not injure the organisms. It would appear, therefore, that the method of Peterkin and Sharpe would permit rapid estimation by measurement of microbial ATP from the cells providing an alternative rapid estimation method for dairy products.

A particular interest at Leatherhead Food Research Association has been the separation of microorganisms by the use of ion exchange chromatography. Most microorganisms are negatively charged at pH

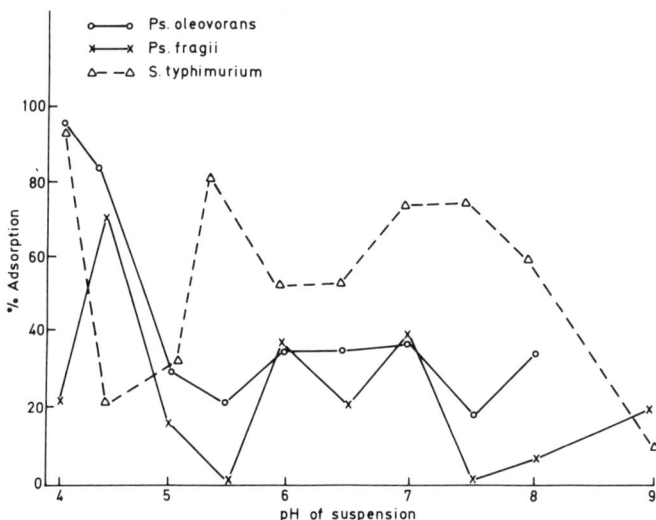

Fig. 14.   Effect of pH on the adsorption of three Gram-negative species of microorganisms to Bio-Rex 70.

values above about 5·0 and will adsorb to positively charged (anion) exchange resins. The behaviour of microorganisms in such systems has been described by Daniels[36] and Wood.[37] The converse reaction, the adsorption of microorganisms to negatively charged (cation) exchange resins at pH values below 5·0 has also been described. The great adsorptive capacities of these materials, 10^{10} cfu g^{-1} of resin, in combination with the wide range of resins available and the simplicity of the procedures provide a wealth of opportunity for the investigation of separation procedures.

In our own work we have made use of the fact that under the correct conditions the Gram-negative bacteria, which form the major part of the microbial flora of raw meat, do not adsorb to cation exchange resin whilst much of the protein from raw meat does so.[38] The basis for this effect can be seen by examining the pH–adsorption spectrum for Gram-negative organisms on this resin (Fig. 14). At pH values of about 5·5 adsorption of these cells to the resin is low. The exploitation of this phenomenon for the separation of microorganisms from meat protein is illustrated in Fig. 15. Raw meat homogenised in sterile buffer (0·05M phosphate–citrate; pH 5·8) and briefly centrifuged to remove coarse particles[38] is passed through a column of cation exchange resin (Bio-Rex 70, Bio-Rad Laboratories) previously equilibrated with the same buffer. Most of the bacteria pass

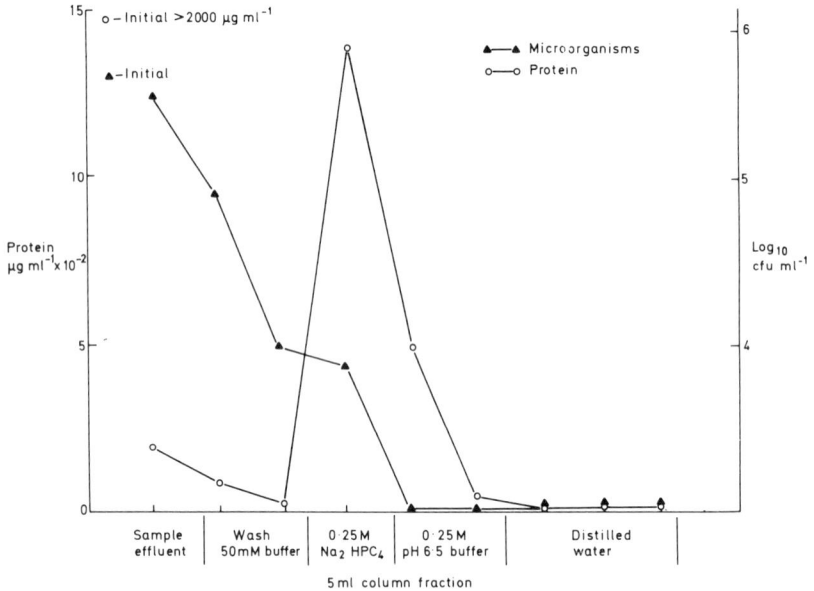

FIG. 15.   Separation of microorganisms from stewing steak.

through the column and are collected in the first two fractions with a small amount of soluble protein from which the cells can be separated by filtration. The recovery of the protein from the resin is achieved by elution with a more concentrated salt solution. The use of a column system has advantages in research work and illustrates well the separation of the microorganisms from meat protein. However, in routine use a batch treatment i.e. simple stirring of the resin with meat supernatant (see Table 2) is more convenient.

The cation exchange resin system may also have potential for the physical separation of different species of microorganisms or for the physical

TABLE 3

THE SEPARATION OF *E. coli* AND *Staph. aureus* BY SELECTIVE ADSORPTION TO CATION EXCHANGE RESIN

	Percentage E. coli	Staph. aureus	*Ratio* E. coli: Staph. aureus
Initial suspension	90	10	9:1
Initial effluent	99·96	0·04	2 490:1
Fractions 8 and 9	4·0	96	1:24

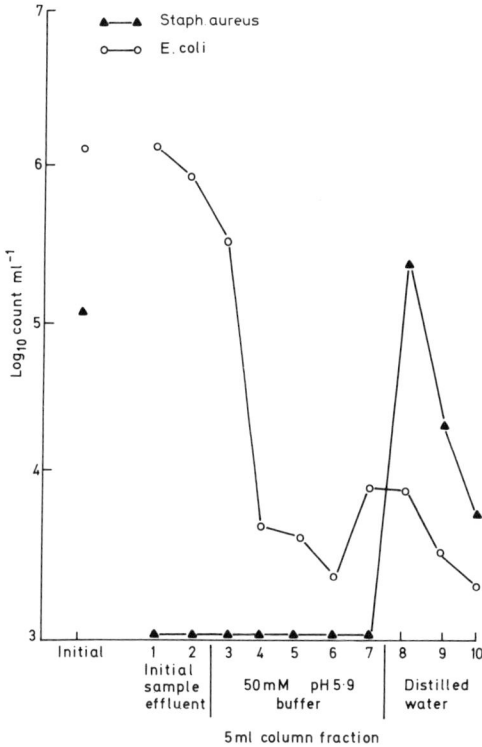

FIG. 16.   Separation of *Staph. aureus* and *E. coli* by differential adsorption to Bio-Rex 70.

enrichment of one species over another. The feasibility of this approach is demonstrated by the differential adsorption of species of bacteria from mixtures of species passed through small columns of resin (Fig. 16). The extent of the separation achieved by this procedure is shown in Table 3.

A considerable enrichment of *Escherichia coli* was evident in the initial effluent from the column compared with the mixture applied to the column. Enrichment of *Staph. aureus* was evident in fractions 8 and 9 i.e. distilled water. Overall, 87 % of the cfu of *E. coli* and 97 % of the cfu of *Staph. aureus* present in the original mixture were recovered in a viable state after the separation process.

Ion exchange resins in all their diversity form only one group of chromatographic adsorbents. In these the possibilities for separation are based upon the electrical charge of the cells.

Another technique which is under investigation is hydrophobic interaction chromatography in which the affinity of a microorganism for a surface (derivatised agarose gels with alkyl or aryl substituents) is dependent upon the hydrophobic/hydrophilic nature of the microbial surface. Wadstrom et al.,[39,40] have used this technique to investigate the surface of enterotoxigenic strains of E. coli and also Dahlbäck has used it in the examination of marine pseudomonads from surface films of sea water.[41] Our own preliminary investigations of this technique at Leatherhead have shown also that food associated organisms, e.g. pseudomonads and lactobacilli, can be manipulated on these gels without loss of viability.

These few observations of ion exchange and hydrophobic interaction chromatography represent empirical approaches to a new field of microbial separation. It seems possible that such investigations will eventually link up with the major and very active fields of work in the chemical structure of the surface layers of microbial cells, e.g. the outer membrane proteins[42] and the physical interaction of microorganisms with solid surfaces.[43] In this case we may come to understand better the basis of the various separation procedures and move to a position where it may be possible to predict the conditions for microbial separations, in much the same way that biochemists do for their materials.

From the viewpoint of the food microbiologist the separation of microorganisms from solid foods may appear to present a daunting prospect, after all a population of 10^6 cfu g^{-1} may represent only 1 ppm of the mass of the food. However, total physical separation is not necessarily required, e.g. in the separation of microbial from non-microbial ATP in foods it may not be necessary to remove all constituents of the food other than ATP unless they produce non-specific interference such as quenching. Similarly, in radiometric or impedimetric estimations, the removal of sufficient of the food constituents so as to permit the concentration of microorganisms from a larger sample of the food would permit more rapid estimations.

In considering the manipulation of microorganisms from foods, the effects of the various procedures on the condition of the organisms is an important consideration. In this context the experiments of Block and Rolland,[44] and Rolland and Block[45] are of interest. These workers used 8 μm Microfiber glass tubes to concentrate salmonellae from river water. The organisms were adsorbed to the glass from water at pH 3·5 and eluted from the glass with beef extract at pH 9·5. The combined effects of exposure of laboratory grown organisms to this rapid variation in pH values was found to vary with the organism; 40 % *S. typhimurium* was inactivated whilst *S. typhi* was not affected. In samples of river water examined, however, *S, typhimurium*, *S. paratyphi* B, *S, panama*, *S. derby* and *S. give* were all concentrated at pH 3·5 even though some of these organisms had been found to be extremely susceptible to this pH under laboratory conditions. This work illustrates two points relevant to the physical manipulation of microorganisms; firstly that whilst pH conditions for optimal growth may be relatively narrow microorganisms can withstand quite harsh conditions for short periods of time. The second point is that we know little about the natural condition of microorganisms. Most of our knowledge of organisms is derived from pure culture work, or from the naturally occurring populations via cultural techniques. It is an interesting thought that the development of physical separation techniques may expand our knowledge of the natural state of microorganisms.

EVALUATION OF NEW METHODS

The investigation of new techniques for the estimation of microbial populations eventually leads to comparisons with the standard method. In food microbiology this most usually means comparison of the new method with the colony count and results in statistical descriptions of the relationship between the methods—correlation, regression, confidence limits. This 'head-on' type of comparison is most suitable for comparing methods which have similar bases of estimation e.g. two colony count techniques. It would appear also to give a straightforward comparison of the direct epifluorescent count against the colony count[4] since the techniques estimate similar quantities i.e. clumps of organisms which might be considered as colony forming units, and colonies.

Comparisons of results obtained by methods which have different bases of estimation e.g. colony count and ATP measurement, are not so straightforward. A problem presented by such a comparison is that the colony forming unit is not a standard quantity in the same sense that a gram of ATP is. Whilst a gram of ATP is, scientifically speaking, a standard amount, the cfu can only be defined as 'that entity which gives rise to a colony' and may vary in quantity even within a single food homogenate. One has only to read the recent work on microbial adhesion[43,46] to appreciate the possibilities for the aggregation and flocculation of organisms which may exist in physically complex systems such as meat homogenates. It would, in fact, be very remarkable to find that the colony forming unit was a uniform quantity under these conditions. This is not intended as a criticism of the colony count (it can equally well be argued that [ATP] may vary from cell to cell) but merely points out a fundamental difference between this method and those that are methods of estimation based on biomass and/or metabolism. The variation in the quantity of the colony forming unit is, of course, incorporated with many other sources of variation e.g. sampling, pipetting and counting errors, into the overall variation observed for this method and is accepted as the norm. Many of these sources of error in colony counts will be common to both types of techniques but variation in the quantity of a colony forming unit will not be and this difference will affect the relationship of the techniques. For example, an ideal relationship between estimates of population obtained by measurement of ATP and by colony count would necessitate a constant ratio of ATP/cfu. This implies that cfu's must contain the same number of microbial cells—an unlikely circumstance.

The main point to be made here is that new methods of estimation e.g. ATP, impedance and radiometry, can be expected to bear a close relationship but only under very particular circumstances would a precise relationship to the colony count be expected. The closeness of these relationships is under active investigation and appears sufficient for microbiologists to avail themselves of the advantages of speed and labour saving. This is as far as such comparisons will allow. Which of the methods provides the more useful information about the food is far from clear at present.

FUTURE DEVELOPMENTS

The preceding discussion begins from the viewpoint that microbiologists wish to estimate the number of colony forming units of microorganisms in foods and considers how other methods e.g. impedance, ATP might provide a measure of colony forming units. This is what is currently happening in food microbiology and it is probable that in the first instance these new methods will be used merely as a rapid index of colony count. However, in the broader view, colony counts are not an end in themselves. They are merely scientific observations which, together with our experience of the system under study, permit some predictions about the 'acceptability' of the food for the consumer or manufacturer. It is, therefore, acceptability which we wish to determine—whether as propensity to spoilage or potential hazard to health—the fact that we have traditionally done so by colony count and presence or absence tests should not limit microbiologists just to these tests. To use impedimetric or radiometric techniques only to determine colony counts is to disregard any other advantage which may be inherent in these types of measurement. Therefore, in the development of new techniques we should attempt, where possible, to evaluate them for their own potential as direct measures of acceptability.

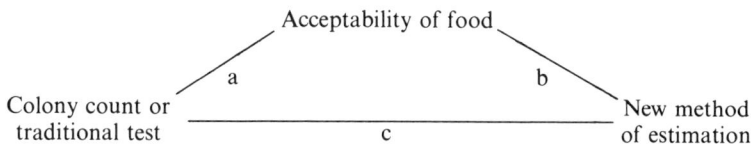

FIG. 17.   Evaluation of a new method of estimation.

In Fig. 17 the relationships 'a' and 'b' are direct and each may have its own particular value as a measure of acceptability. However, to relate the new method to acceptability via 'a' and 'c' automatically restricts the new method to the potential of the traditional method. This illustrates a view of estimation techniques in which the colony count is considered as a single type of scientific observation which can be made and which may be more or even less indicative of the microbiological status of the food than are other types of observation. At present we have more experience in the interpretation of this particular observation than with any other but that does not mean that better methods of predicting the microbiological status of foods cannot be found. A more radical view of food microbiology within the broader field of science has been put forward by Sharpe[47] in *An Alternative to Food Microbiology for the Future*. In this challenging view he questions the traditional concepts used in the microbiological assessment of foods and emphasises the difference between food microbiology and the mainstream of scientific practice.

Sharpe argues that food microbiologists are obsessed with the enumeration of microorganisms and that this obsession has greatly retarded the development of more realistic methods, and in particular instrumented methods, of assessing the microbiological acceptability of foods. His description of the plate count as 'a totally unique datum—nothing in the physical, chemical, biochemical or immunological world corresponds to it and no test based on these properties can ever correspond to it' indicates how far he feels that food microbiology is removed from the major scientific disciplines. His paper goes on to argue that food microbiologists should stop counting 'clusters of bacteria which can grow on agar' and address themselves to the measurement of effects *directly* relevant to acceptability of foods for human consumption i.e. the physical, chemical, biochemical or immunological manifestations of the organisms to which human bodies respond. The general methods by which this could be approached are outlined. This view of food microbiology, whilst highly contentious, represents a development of thought which has been prompted by the desire to use instruments in an area of science which seemed to have been passed-by in the general development of scientific instrumentation.

Whether this view is correct or not is perhaps less important than the fact that it was put forward and represents the desire to investigate beyond the conceptual limitations imposed by our very long standing traditional methodology. It seems reasonably clear that new developments in estimation will be used in routine food microbiology and that initially they will be used to estimate 'colony counts'. This will be an advance in itself; however, it may also be that the use of these techniques will accelerate the development of both methods and concepts within food microbiology.

REFERENCES

1. HEDÉN, C-G., and ILLENI, W. J., *New Approaches to the Identification of Microorganisms*, 1975, John Wiley and Sons, Inc., New York.
2. JOHNSTON, H. H., and NEWSOM, S. W. B., *2nd International Symposium on Rapid Methods and Automation in Microbiology*, 1976, Learned Information (Europe) Ltd, Oxford and New York.
3. PETTIPHER, G. L., and RODRIGUES, U. M., *J. Appl. Bact.*, 1981, **50**, 157.
4. PETTIPHER, G. L., MANSELL, R., McKINNON, C. H., and COUSINS, C. M., *Appl. Environ. Microbiol.*, 1980, **39**, 423.
5. SHARPE, A. N., and KILSBY, D. C., *J. Appl. Bact.*, 1971, **34**, 435.
6. GILCHRIST, J. E., CAMPBELL, J. E., DONELLY, C. B., PEELER, J. T., and DELANEY, J. N., *Appl. Microbiol.*, 1973, **25**, 244.
7. DONELLY, C. B., GILCHRIST, J. E., PEELER, J. T., and CAMPBELL, J. E., *Appl. Environ. Microbiol.*, 1976, **32**, 21.
8. GILCHRIST, J. E., DONELLY, C. B., CAMPBELL, J. E., and PEELER, J. T., *AOAC Abstracts*, 1976, 163a.
9. JARVIS, B., LACH, V. H., and WOOD, J. M., *J. Appl. Bact.*, 1977, **43**, 149.
10. AOAC Method, **56** 1981. Official final action, *Spiral Plate method for bacterial count of foods or cosmetics*, 46.110-46.116 *J. AOAC*, 1981, **64**, 528.
11. RICHARDS, J. C. S., JASON, A. C., HOBBS, G., GIBSON, D. M., and CHRISTIE, R. H., *J. Phys. E: Sci. Instrum.*, 1978, **11**, 560.
12. STEWART, G. N., *J. Exp. Med.*, 1899, **4**, 235.
13. HARDY, D., KRAEGER, S., DUFOUR, S. W., and CADY, P., *Appl. Environ. Microbiol.*, 1977, **34**, 14.
14. CADY, P., HARDY, D., MARTINS, S., DUFOUR, S. W., and KRAEGER, S. J., *J. Food Protection*, 1978, **41**, 277.

15. WOOD, J. M., LACH, V. H., and JARVIS, B., *J. Appl. Bact.*, 1977, **43**, xiv (Abstract).
16. BELL, C., *Food Flavourings, Ingredients, Packaging and Processing*, 1980, **2**, 32.
17. ROWLEY, R. B., PREVITE, J. J., and SRINIVASA, H. P., *J. Food Sci.*, 1978, **43**, 1720.
18. PREVITE, J. J., *Appl. Microbiol.*, 1972, **24**, 535.
19. HATCHER, W. S., DI BENNEDETTO, S., TAYLOR, L. E., and MURDOCK, D. J., *J. Food Sci.*, 1977, **42**, 633.
20. STEWART, B. J., EYLES, M. J., and MURRELL, W. G., *Appl. Environ. Microbiol.*, 1980, **40**, 223.
21. LEVIN, J., and BANG, F. B., *Thrombosis Diathesis and Hemorragica*, 1968, **19**, 186.
22. JORGENSEN, J. H., CARVAJAL, H. F., CHIPS, B. E., and SMITH, R. F., *Appl. Microbiol.*, 1973, **26**, 38.
23. COATES, D. A., *J. Appl. Bact.*, 1977, **42**, 445.
24. JAY, J. M., *J. Appl. Bact.*, 1977, **43**, 99.
25. OLENIACZ, W. D., PISANO, M. A., and ROSENFIELD, D., In *Automation in Analytical Chemistry*, 1966, Technicon Corp. New York.
26. STRANGE, R. E., *Adv. Microbiol. Physiol.*, 1972, **8**, 105.
27. PICCIOLO, G. L., THOMAS, R. R., CHAPPELLE, E. W., TAYLOR, R. E., JEFFERS, E. J., and McGARRY, M. A., *2nd International Symposium on Rapid Methods and Automation in Microbiology*, 1976, eds. Johnston, H. H., and Newson, S. W. B., Learned Information (Europe) Ltd, Oxford and New York.
28. LEVIN, G. V., CHEN, C. S., and DAVIS, E., *Aerospace Med. Res. Lab.*, 1967, TR 67.
29. SHARPE, A. N., WOODROW, M. N., and JACKSON, A. K., *J. Appl. Bact.*, 1970, **33**, 758.
30. BAUMGART, J., FRICKE, K., and HUG, C., *Fleischwirtsch.*, 1980, **60**, 266.
31. JOHNSTON, H. H., MITCHELL, C. J., and CURTISS, G. D. W., In *2nd International Symposium on Rapid Methods and Automation in Microbiology*, 1976, eds. Johnston, H. H., and Newsom, S. W. B. Learned Information (Europe) Ltd, Oxford and New York.
32. VANSTAEN, H., *Lab. Practice*, 1980, 1281.
33. GOLDSCHMIDT, M. C., and FUNG, D. Y. C., *Food Technol.*, 1979, **33**, 63.
34. LUNDIN, A., and THORE, A., *Appl. Microbiol.*, 1975, **30**, 713.
35. PETERKIN, P. I., and SHARPE, A. N., *Appl. Environ. Microbiol.*, 1980, **39**, 1138.
36. DANIELS, S. L., *Dev. Ind. Microbiol.*, 1972, **13**, 211.
37. WOOD, J. M., In *Microbial Adhesion to Surfaces*, eds. Berkeley, R. C. W., Lynch, J. M., Melling, J., Rutter, P. R., and Vincent, B., 1980, Ellis Horwood Ltd, Chichester, England.
38. WOOD, J. M., Ph.D. Thesis, Univ. of Surrey, 1980.
39. WADSTROM, T., SMYTH, C. J., FARIS, A., JOHNSSON, P., and FREER, J. H., *Proc. 2nd Int. Symp. on Neonatal Diarrhoea*, 1978, S. Aires, Soskotoon and Valo.
40. WADSTROM, T., FARIS, A., and HJERTEN, S., In *Microbial Adhesion to Surfaces*, eds. Berkeley, R. C. W., Lynch, J. M., Melling, J., Rutter, P. R., and Vincent, B., 1980, Ellis Horwood, Ltd, Chichester, England.

41. DAHLBÄCK, B., HERMANSSON, M., KJELLEBERG, S., NORKRANS, B., and PEDERSEN, K., in *Microbial Adhesion to Surfaces*, eds. Berkeley, R. C. W., Lynch, J. M., Melling, J., Rutter, P. R., and Vincent, B., 1980, Ellis Horwood Ltd, Chichester, England.
42. OSBOURN, M. J., and WU, H. P. C., *Ann. Rev. Microbiol.*, 1980, **34**, 369.
43. BEACHY, E. H., *Bacterial Adherence Receptors and Recognition*, Series B, Vol. 6, 1980, Chapman and Hall, London and New York.
44. BLOCK, J. C., and ROLLAND, D., *Appl. Environ. Microbiol.*, 1979, **38**, 1.
45. ROLLAND, D., and BLOCK, J. C., *Appl. Environ. Microbiol.*, 1980, **39**, 659.
46. BERKELEY, R. C. W., LYNCH, J. M., MELLING, J., RUTTER, P. R., and VINCENT, B., *Microbial Adhesion to Surfaces*, 1980, Ellis Horwood Ltd, Chichester, England.
47. SHARPE, A. N., *Food Technol.*, 1979, **33**, 71.

INDEX

215